VOLUME ONE HUNDRED AND FIFTY

INTERNATIONAL REVIEW OF NEUROBIOLOGY

Stress and Brain Health: Across the Life Course

INTERNATIONAL REVIEW OF NEUROBIOLOGY

VOLUME 150

VOLUME ONE HUNDRED AND FIFTY

INTERNATIONAL REVIEW OF NEUROBIOLOGY

Stress and Brain Health: Across the Life Course

Edited by

ANGELA CLOW
Psychology, School of Social Sciences, University of Westminster, London, England

NINA SMYTH
Psychology, School of Social Sciences, University of Westminster, London, England

ELSEVIER

ACADEMIC PRESS
An imprint of Elsevier

Academic Press is an imprint of Elsevier
50 Hampshire Street, 5th Floor, Cambridge, MA 02139, United States
525 B Street, Suite 1650, San Diego, CA 92101, United States
The Boulevard, Langford Lane, Kidlington, Oxford OX5 1GB, United Kingdom
125 London Wall, London, EC2Y 5AS, United Kingdom

First edition 2020

ISBN: 978-0-12-816752-6
ISSN: 0074-7742

For information on all Academic Press publications
visit our website at https://www.elsevier.com/books-and-journals

Publisher: Zoe Kruze
Acquisitions Editor: Sam Mahfoudh
Editorial Project Manager: Peter Llewellyn
Production Project Manager: Abdulla Sait
Cover Designer: Greg Harris

Typeset by SPi Global, India

Contents

Contributors

Justin L. Bollinger
Department of Pharmacology & Systems Physiology, University of Cincinnati College of Medicine, Cincinnati, OH, United States

Angela Clow
Psychology, School of Social Sciences, University of Westminster, London, England

Michael C. Craig
Institute of Psychiatry, Psychology and Neuroscience, King's College London; National Female Hormone Clinic, Maudsley Hospital, SLaM NHS Foundation Trust, London, United Kingdom

Alexander J. Dufford
Department of Psychology, University of Denver, Denver, CO, United States

Melissa L. Engel
Department of Psychology, Emory University, Atlanta, GA, United States

Gary W. Evans
Departments of Design and Environmental Analysis and of Human Development, Cornell University, Ithaca, NY, United States

Lourdes Fañanás
Department of Evolutionary Biology, Ecology and Environmental Sciences, Faculty of Biology, University of Barcelona (UB), Barcelona; Centro de Investigación Biomédica en Red en Salud Mental (CIBERSAM), Madrid, Spain

Vivette Glover
Visiting Professor of Perinatal Psychobiology, Imperial College London, London, United Kingdom

Megan R. Gunnar
Institute of Child Development, University of Minnesota, Minneapolis, MN, United States

Akshaya Hegde
School of Biological Sciences, Nanyang Technological University, Singapore, Singapore

Steve Horvath
Department of Human Genetics, David Geffen School of Medicine; Department of Biostatistics, Fielding School of Public Health, University of California Los Angeles, Los Angeles, CA, United States

Pilyoung Kim
Department of Psychology, University of Denver, Denver, CO, United States

Alexandra Lautarescu
Institute of Psychiatry, Psychology and Neuroscience; Perinatal Imaging and Health, Centre for the Developing Brain, School of Biomedical Engineering and Imaging Sciences, King's College London, London, United Kingdom

Robin Law
Psychology, School of Social Sciences, University of Westminster, London, England

Rupshi Mitra
School of Biological Sciences, Nanyang Technological University, Singapore, Singapore

Kelly M. Moench
Department of Psychology, Indiana University-Purdue University, Indianapolis, IN, United States

Helena Palma-Gudiel
Department of Evolutionary Biology, Ecology and Environmental Sciences, Faculty of Biology, University of Barcelona (UB), Barcelona; Centro de Investigación Biomédica en Red en Salud Mental (CIBERSAM), Madrid, Spain

Elena Rossi
Psychology, School of Social Sciences, University of Westminster, London, England

Nina Smyth
Psychology, School of Social Sciences, University of Westminster, London, England

Cara L. Wellman
Department of Psychological & Brain Sciences; Center for the Integrative Study of Animal Behavior; Program in Neuroscience, Indiana University, Bloomington, IN, United States

Carly Wood
School of Life Sciences, University of Westminster, London, England

Anthony S. Zannas
Department of Psychiatry; Department of Genetics; Neuroscience Curriculum, University of North Carolina; Institute for Trauma Recovery, University of North Carolina School of Medicine, Chapel Hill; Department of Psychiatry and Behavioral Sciences, Duke University Medical Center, Durham, NC, United States

CHAPTER ONE

Salivary cortisol as a non-invasive *window on the brain*

Angela Clow*, Nina Smyth
Psychology, School of Social Sciences, University of Westminster, London, England
*Corresponding author: e-mail address: clowa@westminster.ac.uk

Contents

Abstract

The validation of accurate and meaningful assessment of cortisol in saliva samples has proved revolutionary in stress research. Its many advantages have expanded the scope of investigation from traditional laboratory and clinical settings to include multidisciplinary and community-based research. These developments have given rise to a wealth insight into the links between stress and health. Here we highlight the potential of salivary cortisol as both a product and mediator of brain function, instrumental in disturbing brain health. However, the subtleties of salivary cortisol as a measure can be underestimated, leading to misinterpretation of findings. These issues are explored, with a particular emphasis on necessary methodological rigor. Notwithstanding great promise, there is undeniably more to learn so we conclude by making recommendations for future research including use of salivary cortisol in the development of integrative predictive models of stress-related risk factors and resilience across the life course.

International Review of Neurobiology, Volume 150
ISSN 0074-7742
https://doi.org/10.1016/bs.irn.2019.12.003

1. Why is cortisol interesting?

It is increasingly apparent that stress exerts powerful effects on brain health via a complex and highly coordinated stress-response system (Joëls & Baram, 2009; McEwan, 2012; McEwen, Eiland, Hunter, & Miller, 2012). A better understanding of the causative pathways, individual differences in vulnerability and effective individual and population-based counter-measures to stress is important to diminish its harmful effects. Progress in the investigation of these vital issues requires accurate measurement of relevant variables: salivary cortisol is just such a measure. Changing levels of the hormone cortisol in peripheral body fluids are the product of complex brain processes but if measured accurately and interpreted appropriately can provide valuable insight: a non-invasive indirect *window on the brain*. It can be measured not only under laboratory conditions but also in people going about their everyday lives, providing ecological validity. From hour to hour circulating cortisol concentrations can change in response to everyday thoughts and emotions: negative events (e.g., stress) initiate a spike whereas experiences that are more pleasurable cause a reduction. These variations in cortisol secretion can be accurately captured by measurement in saliva. The cortisol response to stress reflects the function of complex brain networks including the amygdala, frontal cortex and hippocampus. At the same time, levels of cortisol exert powerful effects on the brain's structure and function with chronically high levels linked to neurotoxic effects affecting mood, cognition and the stress response itself. In other words, the hormone cortisol is both a product and mediator of brain function, instrumental in disturbing brain health. This bi-directional role underpins the central importance of cortisol in the analysis of stress and brain health. Much research has been dedicated to analyzing the correlates and consequences of the size of the salivary cortisol response to stress and diurnal cortisol patterns in the hope of illuminating biopsychosocial causal pathways, providing a non-invasive biomarker of current brain function and indicator of vulnerability to future brain health. These timely volumes will highlight the role of stress in affecting brain health, interpreting the latest thinking on measurement, mechanisms and potential counter-measures, making them relevant for a wide range of researchers and practitioners seeking insight about future directions and implications for public health.

2. What is cortisol?

Cortisol is an important and pervasive steroid hormone performing a wide range of "housekeeping" duties to ensure healthy functioning. As most bodily cells have cortisol receptors it affects multiple and diverse systems, ranging from regulation of the metabolic, immune, cardiovascular and cognitive systems (McEwan, 2000). All of these functions make cortisol a crucial hormone to protect overall health and well-being. It is the product of a neuroendocrine cascade, meaning it is coordinated from the brain via a signaling system known as the hypothalamic-pituitary-adrenal axis (HPA axis). The HPA axis is a key conduit by which the brain can exert control over physiological activity, which it does in normal everyday activity and also in response to stress. The neural control center for the axis is in the hypothalamus, a region of the brain located below the thalamus within the evolutionary old limbic system, our emotional brain. The paraventricular nucleus (PVN) lies deep within the hypothalamus and is the "trigger point" receiving neuronal input from various modalities including the cognitive and emotional brain (i.e., sensitivity to stressors) as well as the hypothalamic suprachiasmatic nucleus, transmitting environmental information denoting dark-light transitions that informs the circadian pattern of secretion. In response to activation the parvocellular cells of the PVN secrete the neuropeptide, corticotropin releasing factor (CRF) which in turn stimulates the release of adrenocorticotrophic hormone (ACTH) from corticotrophs in the anterior pituitary, an endocrine gland that sits just below the hypothalamus. Adrenocorticotrophic hormone (ACTH) once released into the general circulation stimulates steroidogenic activity and cortisol release from the *zona fasciculate* of the adrenal cortex. Cortisol (corticosterone in rodents) in common with all of the adrenal hormones is derived from the steroid precursor pregnenolone (itself derived from cholesterol).

In summary, the HPA axis is a signaling cascade from the brain to the adrenal cortex, resulting in cortisol secretion into the general circulation. As cortisol is lipid-soluble once in the blood stream it is able to pass freely through cellular plasma membranes, giving it access to all cells, including the brain. Within the brain cortisol's effects are widespread, dependent upon two types of receptor which differ in their distribution and properties: the glucocorticoid receptor (GR) and the mineralocorticoid receptor (MR). Cortisol exerts effects on the brain through both genomic (directly binding to DNA) and non-genomic mechanisms, affecting neurotransmitters, neurotrophic factors, sex hormones and other stress mediators to shape present and future responses to stress (Gray, Kogan, Marrocco, & McEwen, 2017). The MRs are localized in the limbic system and prefrontal cortex, binding cortisol with high affinity, whereas GRs, which are more widely distributed throughout the brain, and bind cortisol with approximately one-tenth of the affinity (de Kloet, Meijer, de Nicola, de Rijk, & Joëls, 2018). MR activation is related to the onset of the stress response whereas GR are associated with facilitation of recovery. The distinct difference in receptor type, distribution and sensitivity allows cortisol to regulate brain function in different ways depending on ambient concentrations: it is the balance between receptor occupancy that determines outcomes (de Kloet, Joëls, & Holsboer, 2005). As reviewed in these volumes the stress response and its feedback is complex, having many and diverse effects on brain structure and function. Salivary cortisol concentrations represent the net effect of these brain processes, underpinning its validity in the study of stress and brain health.

3. Stress and its measurement

Within engineering stress refers to physical strains that lead to structural distortion. This terminology is adopted for living things: biological and psychological disturbance caused by events is known as stress, whereas the stress-inducing events are referred to as stressors. Obviously, there are vital differences between the engineering concept of stress and human stress. Not least of these is the predictability of the relationship between stressor and stress. The engineer can have confidence in a foreseeable relationship between stressor and stress; one steel girder will behave much like the next. This is not the case for human stress where the stressful experience is not directly proportionate to the degree of stressor. For example, a relatively minor stressor might have a large effect on biology and behavior; alternatively, potent external stressors might not necessarily induce proportional stressful reactions, thus human stress is subjective. Psychologists have been studying the cause of these individual differences for years and emphasize the mediating role of appraisal, coping, and emotion: depending on goals, beliefs, and coping strategies, a stressor may be appraised differently and elicit different stress reactions between and within individuals. A large part of the salivary cortisol literature deals with this issue including the role of pre-natal, early life development and environmental influences that can affect vulnerability and resilience to stressors, all of which are discussed in this volume. In addition, our reactions to stress can change across the lifespan and there are things we, as individuals and communities, can do to counteract the negative effects of stressors on our systems.

Stress can be defined and objectively measured in terms of its biological response. This response is composed of two distinct but interacting physiological systems: the sympathetic nervous system (SNS) with noradrenaline as the key mediator, and the HPA axis with cortisol as the main downstream hormone. Whilst the SNS is not uniquely activated by threat, i.e., stress (it is also activated by excitement and other forms of positive arousal) activation of the HPA axis is fine-tuned to react to threat and is therefore a "cleaner" indicator of stress than study of the SNS alone. Together the stress-response systems allow for adjustment to a wide range of bodily and environmental demands and in this way are adaptive and beneficial for outcomes. Stress-related problems for the brain and body arise when stress-reactions are persistent and sustained—known as "chronic stress." Accordingly, in the short-term stress response systems are adaptive but in the long-term, under sustained activation they become maladaptive, leading to ill-health.

It is interesting to consider that the stress-response to psychosocial threat is equivalent to that to physical threat. So, the stress-response machinery that facilitates the appropriate energy for the flight and fight response (necessary if we are confronted by an assailant, for example) is totally out of proportion for dealing with common psychosocial stress (e.g., difficult relationships, inadequate housing, pressure and lack of control at work). Standardized exposure to short-term psychosocial stress in the laboratory increases salivary cortisol levels (Dickerson & Kemeny, 2004). It is very interesting to examine the most reliable stimuli for generating substantial stress response are threat to self-esteem, lack of control and novelty. Sadly, today exposure to these stimuli can be frequent everyday events that can result in harmful repeated stress responding, i.e., chronic stress, and ultimately affect brain function and health outcomes. Of course, the experience of stress can affect broader function than these two response systems. For example, behavior is usually affected, often leading to maladaptive stress coping mechanisms such as smoking and drinking alcohol. Sleep and appetite are frequently disrupted whilst mood, memory and thinking styles can be modulated. These additional stress responses can also be measured, but for the purposes of studying stress and brain health the primary focus of investigation is the role of the HPA axis as the primary instrument of change.

4. Salivary cortisol measurement in human studies

For many years cortisol was typically only measured in humans from single measure blood samples or 24-h urine collections for use in clinical diagnostics. Clemens Kirschbaum, working in the celebrated biological psychology research group of Dirk Hellhammer in Germany first validated accurate and meaningful assessment of cortisol in saliva samples (see Kirschbaum & Hellhammer, 1989). This step proved revolutionary due to its many advantages over blood and urine, taking investigation of cortisol secretion away from the clinical setting in to a realm of multidisciplinary research. It opened up opportunities to investigate the dynamically changing state of biologically relevant HPA axis activity from repeated sampling of easily accessed saliva. In the bloodstream, most cortisol (about 90%) is stored as a reservoir bound to specific binding proteins transcortin and albumin. Once bound in this way it is transported around the body but is not biologically active. Only unbound cortisol is 'free' as it is available for receptor binding making it biologically active. Assessment of cortisol in blood samples measures *total* cortisol with no distinction between bound and free. Without accurate knowledge of how much is bound to the proteins, it is impossible to estimate the biologically active component without undertaking the complicated process of osmosis, which separates the two components in a test tube. In contrast, assessment of salivary cortisol measures *only* the biologically active "free" component of cortisol as the bound component is too large a molecule to pass through the saliva glands. Another substantial advantage of using saliva as the medium in which to measure cortisol is that passage through the saliva glands is passive, not an active transport system, as for salivary IgA, for example. This means that the concentration of cortisol in saliva is not dependent on salivary volume, in other words secretion does not have to be calculated per unit time, which would make it methodologically much more challenging. Any volume of saliva collected over any length of time (so long as the volume is adequate for the assay) can generate accurate measures of biologically active cortisol, representative of that circulating through the entire body at any one time. These advantages together with sampling convenience and ease of repeated sampling, even outside of research settings provides huge opportunities for neuroendocrine research.

5. Stress reactivity studies

A commonly used approach is to examine the scale of cortisol responding following exposure to a standardized psychosocial stressor, such as the Trier Social Stress Test (Kirschbaum, Pirke, & Hellhammer, 1993). Responses interact with a wide range of demographic and psychosocial variables (Kudielka, Hellhammer, & Wust, 2009; Zänkert, Bellingrath, Wüst, & Kudielka, 2018) making it problematic to associate different cortisol responses with specific patterns of brain activation (Dedovic, Duchesne, Andrews, Engert, & Pruessner, 2009). It is interesting however that hyper- and hypo-responding of the HPA axis are associated with distinct negative future health outcomes (Turner et al., 2019) suggesting they are associated with distinct patterns of brain activation. Exaggerated reactivity predicts greater risk of coronary artery calcification, hypertension and more rapid telomere attrition. In contrast, blunted HPA axis reactivity predicts more depressive and PTSD symptomatology, greater musculoskeletal pain and lower bone mineral content (Turner et al., 2019). These findings, implicating both exaggerated and blunted cortisol responding to stress, are problematic as there are no guidelines for the optimal "or goldilocks" stress response (i.e., not to large or too small). This complication reinforces the need to be vigilant to include appropriately matched healthy control groups from which to compare populations of interest.

Blunted stress-induced salivary cortisol secretion is an interesting index of dysregulated HPA axis function as, amongst other things; it has been associated with neuroticism and impulsivity, which are characteristic of various behavioral disorders. Although it remains a developing story, it appears to be an indicator of Reward Deficiency Syndrome, reflecting fronto-limbic dysregulation (Carroll, Ginty, Whittaker, Lovallo, & de Rooij, 2017). Optimal responses to stress occur within a normal range (the "goldilocks" zone) and deviation in either direction signals poor systems integration. Adequate cortisol stress reactivity is purportedly protective for brain health via mediation of reward system activity (Oei, Both, van Heemst, & van der Grond, 2014), which is consistent with blunted cortisol reactivity to stress in clinical depression (Burke, Davis, Otte, & Mohr, 2005). Of course, studies of stress reactivity with peripheral markers, such as salivary cortisol, cannot pinpoint the precise site of brain dysregulation, but they can provide useful insight by revealing a deviation from the 'norm' worthy of further investigation.

This area of research is also complicated by the issue of non-responding. Within any experimental stress reactivity study a notable percentage of participants (e.g., 35%, Smyth, Flynn, et al., 2019) will not mount a salivary cortisol response, defined as either a 1.5 nmol/L or 15.5% percentage baseline-to-peak increase (Miller, Plessow, Kirschbaum, & Stalder, 2013). A lower criteria (i.e., 1 nmol/L) might apply depending on the assay used for cortisol determination (see Miller et al., 2013). The issue of non-responding has been relatively ignored but in our data of healthy females we have not observed any association between demographic or psychosocial variables and non-responding (Smyth, Flynn, et al., 2019). Incidence of non-responding in males, who typically exhibit larger responses (Herbison et al., 2016), has not been reported. Non-responding is assumed to be random, attributed to the timing of the stressor coinciding with the refractory phase of the individual underlying ultradian rhythm. Rodent studies have demonstrated that if stress exposure coincides with the falling phase of the ultradian burst both the behavioral and neuroendocrine response to stress is markedly attenuated (Lightman & Conway-Campbell, 2010; Sarabdjitsingh et al., 2010), although this has not been examined in humans. As non-responding is suggested to be random, not a form of blunted HPA axis reactivity, it is recommended to present data analyses with and without their inclusion (Miller et al., 2013). In addition, the prevalence and characteristics of non-responders should be reported to inform data on their characteristics and enable transparent monitoring of data handling.

6. Basal pattern of salivary cortisol secretion

The basal patterns of cortisol secretion reveals insight about brain processes involved in circadian regulation, which are vital for brain health (Karatsoreosa et al., 2011; Oster et al., 2016). However, it has proved surprisingly tricky to tease out the role of stress-related cortisol secretion in brain health precisely because of the changing levels of cortisol over the day. Cortisol secretion is subject not only to perceptions of stress (i.e., stress-responsive) but also to the brain's internal clock, which synchronizes a characteristic 24-h (circadian) pattern. This input from the suprachiasmatic nucleus provides the basal platform of cortisol secretion on which stress reactivity is superimposed, and sustained stress over a period of time ultimately dysregulates this cycle. Healthy salivary cortisol concentrations change about 20-fold per day/night cycle, providing an important chemical signal to downstream processes about appropriate activity for the time of day, and it is this that makes cortisol essential for daily functioning. The cortisol circadian cycle acts as the conductor of a physiological orchestra ensuring disparate body systems remain aligned for maximum efficiency and healthy flourishing of the whole being, i.e., integrated healthy functioning. There are beneficially high levels of cortisol after awakening, known as the cortisol awakening response (CAR; Pruessner et al., 1997) and a decline from morning (about 30 min post-awakening) until bedtime, known as the diurnal decline and lowest levels during the early phases of sleep. Unfortunately, the significance of this circadian cycle was overlooked in many early studies, compromising experimental rigor. It is now apparent however that measurement of this methodologically problematic circadian cortisol cycle in saliva is itself a potent *window on the brain*, as its disruption by chronic stress is an early indicator of dysregulated brain function and the major route by which stress affects brain health (Adam et al., 2017; Gianaros et al., 2017). Aberrant patterns of cortisol secretion may not always translate into concurrent observable symptoms but the reason they are interesting is that, if sustained for any length of time, they provide evidence of causality and a useful early indicator of negative future health outcomes. As such, study of this hormone in saliva samples provides a useful pre-clinical indicator of the links between mind and body in currently healthy and clinical populations (Smyth, Hucklebridge, Thorn, Evans, & Clow, 2013).

7. The ultradian pattern of cortisol secretion and the cortisol awakening response

Cortisol secretion is dynamic, subject to negative feedback that induces an oscillating secretory pattern, with short-lived secretory bursts followed by intra-pulse intervals of approximately 1 h. This means that HPA axis function is pulsatile with the resultant pattern of cortisol secretion known as the ultradian rhythm. The amplitude (to a less extent the rate) of pulses varies throughout the day, with a peak after awakening and a trough at sleep onset, providing the basis of the overarching circadian pattern. Hourly bursts of cortisol secretion are evident in peripheral body fluids of individuals when using frequent (or continuous) blood sampling. They are not apparent in less frequent sampling protocols and when data from multiple individuals is summed to reveal overall trends. The importance of pulsatility in HPA axis function is attracting increasing attention with the pulses (rather than overall concentrations) shown to influence GR-mediated behavioral and neuroendocrine responses to stress, circadian clock genes, glutamatergic transmission and synaptic plasticity in the hippocampus (Fitzsimons et al., 2016; Flynn, Conway-Campbell, Stafford, & Lightman, 2018). It is not surprising therefore, that disruption of the pulsatile ultradian rhythm is implicated in a range of neuropathology (Fitzsimons et al., 2016). Evidently it is important to be aware of this interesting and important area; however, analysis of hormone pulsatility is not easy in saliva samples as the individuals within group data will have pulses at slightly different times resulting in the mean concentration removing evidence of the underlying ultradian rhythm. More recently, interest has focused upon the cortisol awakening response (CAR) as the first pulsatile event of the day (Evans, Smyth, Thorn, Hucklebridge, & Clow, 2019).

The CAR is usually measured from saliva samples and is necessarily measured relative to the moment of awakening, a time described as the "tipping point" of the day, i.e., the moment when the brain switches from night-time sleep to daytime consciousness (Clow, Hucklebridge, Stalder, Evans, & Thorn, 2010). The process of awakening is thought to synchronize the start of the ultradian rhythm so that the CAR, even measured in saliva in multiple participants, is apparent as the coordinated first pulse of the day. Later in the day (more remote from the awakening tipping point), even small inter-individual differences in ultradian timing smooth out all evidence of pulsatility. The underlying ultradian rhythm might also be why some people do not show a CAR, i.e., awakening during a refractory phase (as for non-responding in stress reactivity studies, discussed above). We have found evidence of this in nearly 20% of days despite accurate sampling (Smyth, Clow, Hucklebridge, Thorn, & Evans, 2013); however, no study has examined if this non-CAR is a "trait" characteristic. Studies measuring CAR on repeated days (up to a week) are required to examine this.

The CAR is the only time in the circadian cycle when cortisol levels rise peaking at around 30–45 min post awakening. This means that study of the CAR, from saliva sampling, may provide even more insight into brain function than the rest of the diurnal cycle as it provides insight into not just an index of circadian but also ultradian function. Consistent with this it has recently been proposed that CAR salience, i.e., the shape, including the rate of cortisol decline following the peak is a key measure, indicative of the salience of the underlying ultradian pulse (Evans et al., 2019). This is a unique feature of the CAR and may account for its sensitivity as a biomarker in stress research (Clow et al., 2010). It is proposed that the CAR (as an ultradian marker) has a particular role in synchronization of peripheral (or "slave") clocks throughout the body (Clow et al., 2010). It is clear that, stress-related flat diurnal declines are associated a wide range of physical and mental health (Adam et al., 2017). Similarly, stress-associated flattening of the CAR has been shown to be associated with reduced hippocampal integrity and volume and impaired brain function (Clow et al., 2014; Law & Clow, 2020; Shi et al., 2018). It is proposed that this is due to compromised synchronization between the central clock in the hypothalamic suprachiasmatic nucleus (via the CAR) and slave clocks in the peripheral tissue, including brain (Clow et al., 2010). Such lack of synchronization is known to impact brain health (Menet & Rosbash, 2011). Certainly, this intriguing possibility needs further exploration as if true provides a powerful tool in neuroendocrine research and a potential target for intervention for a wide range of brain conditions.

8. Alternative strategies to probe brain function using salivary cortisol

Of course, measures of salivary cortisol are limited; they do not provide a literal picture of brain function. Instead, if judiciously applied and interpreted, these measures provide valuable *indirect* clues of current brain function and an early warning signal of possible forthcoming neurotoxic effects. However, some strategies can probe more deeply. For example, the dexamethasone suppression test is a tried and tested way of revealing brain cortisol receptor sensitivity. Non-suppression of salivary cortisol concentrations indicates that the administered glucocorticoid was unable to shut down the HPA axis generation of new cortisol, and this is a finding prevalent in clinical depression (Arana, Baldessarini, & Ornsteen, 1985; Carroll, 1982). Similarly, metyrapone is used to probe sensitivity of the HPA axis by temporarily inhibiting the hydroxylation of 11-deoxycortisol into cortisol, effectively blocking cortisol biosynthesis. Such an approach has been used, alongside salivary cortisol monitoring to explore the role of the CAR in memory retrieval (Rimmele, Meier, Lange, & Born, 2010).

To date the primary focus of salivary cortisol research has been to inform causative pathways linking stress and brain health. Opportunities now exist for applying this knowledge to evaluate stress-reduction strategies, at the individual and community level. As reviewed in these volumes much promising intervention work has been undertaken but unfortunately evaluation strategies can be methodologically flawed (Smyth, Rossi, & Wood, 2020). Introduction of simple measures of basal levels of salivary cortisol (if properly collected, see below) can provide valuable objective insight into change, even in clinical conditions (e.g., Smyth, Skender, et al., 2019). Such evidence lends weight to the effectiveness, or not, of interventions—not just as a remote biomarker but as a significant instrument of changing brain health. The important message here is that measures of salivary cortisol are responsive to changing environmental and individual circumstances and that these changes reflect changing brain health in terms of HPA axis status and potential neurotoxic effects.

9. Methodological issues when measuring basal salivary cortisol secretion

The multiple advantages of using saliva as a sampling medium from which to measure cortisol secretion have made it overwhelmingly popular in multidisciplinary research. However, the advantages also bring methodological challenges. The primary issue is around the marked circadian nature of basal cortisol secretion. Self-collection of samples has opened up research into areas not previously possible and been able to capture daily life, with ecological validity. The simplicity has enabled sampling from virtually any participant group from tiny infants to diverse and disabling clinical conditions. Such studies typically seek to capture a picture of the daily (diurnal) pattern of cortisol secretion asking for repeated sampling at different times across the day. The moment of morning awakening, i.e., the transition from sleep to consciousness, is the starting point in each 24-h cycle. It begins with a synchronized surge in cortisol secretion reflecting the first ultradian pulse of the day (i.e., the CAR). This pattern means that all assessment of cortisol secretion should be synchronized to the individual time of awakening. It is not adequate to use clock time, as there are large individual differences in awakening time (Edwards, Clow, Evans, & Hucklebridge, 2001). This means that accurate timing of sampling when using self-collection study designs is of critical importance for measurement of morning levels.

Accurate assessment of the CAR has proved to be the most methodologically problematic area in salivary cortisol research. It is demanding of study participants to collect samples immediately on awakening and at 15 min intervals up to 30–45 min post-awakening (without eating or having a cup of tea/coffee). To successfully capture the moment of awakening and self-collect saliva samples at the correct time intervals and in the correct tubes is difficult, especially when suffering from considerable sleep inertia. Sampling inaccuracy leads to erroneous measurement of the CAR (Smyth, Clow, et al., 2013). However, errors can be minimized if electronic measurement of the sampling process are used (Smyth, Thorn, Hucklebridge, Clow, & Evans, 2016). The CAR is also subject to considerable day-to day variability, in response to state variation (Law, Hucklebridge, Thorn, Evans, & Clow, 2013) requiring measurement on repeated days in order to assess an average, i.e., trait-like CAR

(Hellhammer et al., 2007). These issues have led to much debate in the area leading to expert consensus guidelines that are *essential* reading for all potential researchers of this particular aspect of cortisol section (Stalder et al., 2016). The effort involved in CAR measurement is considerable. If it is not possible to follow the published guidelines it is best to focus upon measurement of the diurnal decline as a measure of basal cortisol secretion (Adam et al., 2017). This approach should avoid inclusion of the CAR period and sample between 3 and 6 times over the post-awakening day, up until the evening.

10. Conclusions

This chapter highlights the potential of using measures of salivary cortisol as a bi-directional stress-related indicator of brain health. As reviewed in these two volumes cortisol is both the product of a stress response system and capable of having profound effects on brain structure and function. Study of the developmental and psychosocial predictors of stress responding can provide some insight into the causal origins of such dysregulated brain function. Similarly, examination of the underlying basal circadian pattern of cortisol secretion, especially the CAR, sheds a light on how well this vital neuroendocrine system is working. Efficient circadian/ultradian systems are essential for health, regulating disparate physiological processes, including brain function. Flattening of the circadian pattern of cortisol secretion is an indicator of disrupted circadian function and can be detected even before signs of aberrant brain function appear, providing the opportunity for monitoring and early intervention. Additionally, the sensitivity of salivary cortisol secretion to stress-alleviating interventions makes it a valuable marker of effectiveness.

Clearly, use of salivary cortisol in the investigation of brain health has great potential. However, there is undeniably a lot more to learn in order to interpret different cortisol profiles in relation to specific indices of brain health. Crucially, the area suffers from a dearth of work using salivary cortisol in large-scale population based prospective studies to test its value as a mediator between stress and brain health. A recent meta-analysis reported just 10% of published studies used a longitudinal methodology (Adam et al., 2017). Use of salivary cortisol measures alongside a range of brain health/behavioral outcomes, multimodal neuroimaging and network neuroscience in both males and females would also do much to inform and facilitate future utility. A vital aim must be to build integrative predictive models (Stringer & Tommerdahl, 2015) to identify stress-related risk factors across the life course, as well as protective factors. Inclusion of salivary cortisol in such models would provide an accessible and valuable adjunct to the study of brain health and resilience.

References

Adam, E. K., Quinn, M. E., Tavernierd, R., Mollie, T., McQuillane, M. T., Dahlkef, K. A., et al. (2017). Diurnal cortisol slopes and mental and physical health outcomes: A systematic review and meta-analysis. *Psychoneuroendocrinology*, *83*, 25–41.

Arana, G. W., Baldessarini, R. J., & Ornsteen, M. (1985). The dexamethasone suppression test for diagnosis and prognosis in psychiatry. Commentary and review. *Archives of General Psychiatry*, *42*, 1193–1204.

Burke, H. M., Davis, M. C., Otte, C., & Mohr, D. C. (2005). Depression and cortisol responses to psychological stress: A meta-analysis. *Psychoneuroendocrinology*, *30*, 846–856.

Carroll, B. J. (1982). The dexamethasone suppression test for melancholia. *British Journal of Psychiatry*, *140*, 292–304.

Carroll, D., Ginty, A. T., Whittaker, A. C., Lovallo, W. R., & de Rooij, S. R. (2017). The behavioural, cognitive, and neural corollaries of blunted cardiovascular and cortisol reactions to acute psychological stress. *Neuroscience and Biobehavioural Reviews*, *77*, 74–86.

Clow, A., Hucklebridge, F., Stalder, T., Evans, P., & Thorn, L. (2010). The cortisol awakening response: More than a measure of HPA axis function. *Neuroscience and Biobehavioural Reviews*, *35*, 97–103.

Clow, A., Law, R., Evans, P., Vallence, A. M., Hodyl, N. A., Goldsworthy, M. R., et al. (2014). Day differences in the cortisol awakening response predict day differences in synaptic plasticity in the brain. *Stress*, *17*, 219–223.

de Kloet, E. R., Joëls, M., & Holsboer, F. (2005). Stress and the brain: From adaptation to disease. *Nature Reviews Neuroscience*, *6*, 463–475.

de Kloet, E. R., Meijer, O. C., de Nicola, A. F., de Rijk, R. H., & Joëls, M. (2018). Importance of the brain corticosteroid receptor balance in metaplasticity, cognitive performance and neuro-inflammation. *Frontiers in Neuroendocrinology*, *49*, 124–145. https://doi.org/10.1016/j.yfrne.2018.02.003.

Dedovic, K., Duchesne, A., Andrews, J., Engert, V., & Pruessner, J. C. (2009). The brain and the stress axis: The neural correlates of cortisol regulation in response to stress. *NeuroImage*, *47*, 864–871.

Dickerson, S. S., & Kemeny, M. E. (2004). Acute stressors and cortisol responses: A theoretical integration and synthesis of laboratory research. *Psychological Bulletin*, *130*, 355–391.

Edwards, S., Clow, A., Evans, P., & Hucklebridge, F. (2001). Exploration of the awakening cortisol response in relation to diurnal cortisol secretory activity. *Life Sciences*, *68*, 2093–2103.

Evans, P., Smyth, N., Thorn, L., Hucklebridge, F., & Clow, A. (2019). Salience versus magnitude in the measurement of the cortisol awakening response. *Psychoneuroendocrinlogy*, *103*, 249–258.

Fitzsimons, C. P., Herbert, J., Schouten, M., Meijer, O. C., Lucassen, P. L., & Lightman, S. (2016). Circadian and ultradian glucocorticoid rhythmicity: Implications for the effects of glucocorticoids on neural stem cells and adult hippocampal neurogenesis. *Frontiers in Neuroendocrinology*, *41*, 44–58.

Flynn, B. P., Conway-Campbell, B. L., Stafford, L., & Lightman, S. L. (2018). The emerging importance of ultradian glucocorticoid rhythms within metabolic pathology. *Annals of Endocrinology*, *79*, 112–114.

Gianaros, P. J., Kuan, D. C.-H., Marsland, A. L., Sheu, L. K., Hackman, D. A., Miller, K. G., et al. (2017). Community socioeconomic disadvantage in midlife relates to cortical morphology via neuroendocrine and cardiometabolic pathways. *Cerebral Cortex*, *27*, 460–473.

Gray, J. D., Kogan, J. F., Marrocco, J., & McEwen, B. S. (2017). Genomic and epigenomic mechanisms of glucocorticoids in the brain. *Nature Reviews Endocrinology*, *13*, 661–673.

Hellhammer, J., Fries, E., Schweisthal, O. W., Schlotz, W., Stone, A. A., & Hagemann, D. (2007). Several daily measurements are necessary to reliably assess the cortisol rise after awakening: State- and trait components. *Psychoneuroendocrinology*, *32*, 80–86.

Herbison, C. E., Henley, D., Marsh, J., Atkinson, H., Newnham, J. P., Matthews, S. G., et al. (2016). Characterization and novel analyses of acute stress response patterns in a population-based cohort of young adults: Influence of gender, smoking, and BMI. *Stress*, *19*(2), 139–150. https://doi.org/10.3109/10253890.2016.1146672.

Joëls, M., & Baram, T. Z. (2009). The neuro-symphony of stress. *Nature Reviews Neuroscience*, *10*, 459–466.

Karatsoreosa, I. N., Bhagat, S., Bloss, E. B., Morrison, J. H., Bruce, S., & McEwen, B. S. (2011). Disruption of circadian clocks has ramifications for metabolism, brain, and behaviour. *Proceedings of the National Academy of Sciences*, *108*, 1657–1662.

Kirschbaum, C., & Hellhammer, D. H. (1989). Salivary cortisol in psychobiological research: An overview. *Neuropsychobiology*, *22*, 150–169.

Kirschbaum, C., Pirke, K. M., & Hellhammer, D. H. (1993). The Trier Social Stress Test—A tool for investigating psychobiological stress responses in a laboratory setting. *Neuropsychobiology*, *28*, 76–81.

Kudielka, B. M., Hellhammer, D. H., & Wust, S. (2009). Why do we respond so differently? Reviewing determinants of human salivary cortisol responses to challenge. *Psychoneuroendocrinology*, *34*, 2–18.

Law, R., & Clow, A. (2020). Stress, the cortisol awakening response and cognitive function. In A. Clow & N. Smyth (Eds.), *Vol. 150. International Review of Neurobiology* (pp. 187–217). Elsevier.

Law, R., Hucklebridge, F., Thorn, L., Evans, P., & Clow, A. (2013). State variation in the cortisol awakening response. *Stress*, *16*, 483–492.

Lightman, S. L., & Conway-Campbell, B. L. (2010). The crucial role of pulsatile activity of the HPA axis for continuous dynamic equilibration. *Nature Review Neuroscience*, *11*, 710–718.

McEwan, B. S. (2000). Allostasis and allostatic load: Implications for neuropsychopharmacology. *Neuropsychopharmacology*, *22*, 108–124.

McEwan, B. S. (2012). Brain on stress: How the social environment gets under the skin. *Proceedings of the National Academy of Sciences of the United States of America*, *109*, 17180–17185.

McEwen, B. S., Eiland, L., Hunter, R. G., & Miller, M. M. (2012). Stress and anxiety: Structural plasticity and epigenetic regulation as a consequence of stress. *Neuropharmacology*, *62*, 3–12.

Menet, J. S., & Rosbash, M. (2011). When brain clocks lose track of time: Cause or consequence of neuropsychiatric disorders. *Current Opinion in Neurobiology*, *21*, 849–857.

Miller, R., Plessow, F., Kirschbaum, C., & Stalder, T. (2013). Classification criteria for distinguishing cortisol responders from nonresponders to psychosocial stress: Evaluation of salivary cortisol pulse detection in panel designs. *Psychosomatic Medicine*, *75*, 832–840.

Oei, N. Y. L., Both, S., van Heemst, D., & van der Grond, J. (2014). Acute stress-induced cortisol elevations mediate reward system activity during subconscious processing of sexual stimuli. *Psychoneuroendocrinology*, *39*, 111–120.

Oster, H., Challet, E., Ott, V., Arvat, E., de Kloet, E. R., Dijk, D. J., et al. (2016). The functional and clinical significance of the 24-hour rhythm of circulating glucocorticoids. *Endocrinology Review, 38*, 3–45.

Pruessner, J. C., Wolf, O. T., Hellhammer, D. H., Buske-Kirschbaum, A., Von Auer, K., Jobst, S., et al. (1997). Free cortisol levels after awakening: A reliable biological marker for the assessment of adrenocortical activity. *Life Sciences, 61*(26), 2539–2549.

Rimmele, U., Meier, F., Lange, T., & Born, J. (2010). Suppressing the morning rise in cortisol impairs free recall. *Learning and Memory, 17*, 186–190.

Sarabdjitsingh, R. A., Conway-Campbell, B. L., Leggett, J. D., Waite, E. J., Meijer, O. C., de Kloet, E. R., et al. (2010). Stress responsiveness varies over the ultradian glucocorticoid cycle in a brain-region-specific manner. *Endocrinology, 151*, 5369–5379.

Shi, X., Sun, X., Yao, Z., Yuan, Y., Wu, J., & Clow, A. (2018). The cortisol awakening response predicts response inhibition in the afternoon of the same day. *Psychoneuroendocrinology, 89*, 23–29.

Smyth, N., Clow, A., Hucklebridge, F., Thorn, L., & Evans, P. (2013). Delays of 5–15 minutes between awakening and the start of saliva sampling matter in assessment of the cortisol awakening response. *Psychoneuroendocrinology, 38*, 1476–1483.

Smyth, N., Flynn, M., Rajcani, J., Hucklebridge, F., Thorn, L., Wood, C., et al. (2019). Attenuated cortisol reactivity to psychosocial stress is associated with greater visual dependency in postural control. *Psychoneuroendocrinology, 104*, 185–190.

Smyth, N., Hucklebridge, F., Thorn, L., Evans, P., & Clow, A. (2013). Salivary cortisol as a biomarker in social science research. *Social and Personality Psychology Compass*, 7, 605–625.

Smyth, N., Rossi, E., & Wood, C. (2020). Effectiveness of stress-relieving strategies in regulating patterns of cortisol secretion and promoting brain health. In A. Clow & N. Smyth (Eds.), *Vol. 150. International Review of Neurobiology* (pp. 219–246). Elsevier.

Smyth, N., Skender, E., Munoz, M., David, F., Fantuzzi, G., Clow, A., et al. (2019). Endurance exercise reduces cortisol in Parkinson's disease with mild cognitive impairment. *Movement Disorders, 34*, 1238–1239.

Smyth, N., Thorn, L., Hucklebridge, F., Clow, A., & Evans, P. (2016). Assessment of the cortisol awakening response: Real-time analysis and curvilinear effects of sample timing inaccuracy. *Psychoneuroendocrinology*, *74*, 380–386.

Stalder, T., Kirschbaum, C., Kudielka, B. M., Adam, E. K., Pruessner, J. C., Wüst, S., et al. (2016). Assessment of the cortisol awakening response: Expert consensus guidelines. *Psychoneuroendocrinology, 63*, 414–432.

Stringer, S., & Tommerdahl, J. (2015). Building bridges between neuroscience, cognition and education with predictive modeling. *Mind, Brain, and Education, 9*(2), 121–126.

Turner, A., Smyth, N., Hall, S., Torres, S., Hussein, M., Jayasinghe, S., et al., (2019). Psychological stress reactivity and future health and disease outcomes: A systematic review of prospective evidence, *Psychoneuroendocrinology*, in press.

Zänkert, S., Bellingrath, S., Wüst, S., & Kudielka, B. M. (2018). HPA axis responses to psychological challenge linking stress and disease: What do we know on sources of intra-and interindividual variability? *Psychoneuroendocrinology, 105*, 86–97.

CHAPTER TWO

Prenatal stress: Effects on fetal and child brain development

Alexandra Lautarescu[a,b], Michael C. Craig[a,c], Vivette Glover[d,*]
[a]Institute of Psychiatry, Psychology and Neuroscience, King's College London, London, United Kingdom
[b]Perinatal Imaging and Health, Centre for the Developing Brain, School of Biomedical Engineering and Imaging Sciences, King's College London, London, United Kingdom
[c]National Female Hormone Clinic, Maudsley Hospital, SLaM NHS Foundation Trust, London, United Kingdom
[d]Visiting Professor of Perinatal Psychobiology, Imperial College London, London, United Kingdom
*Corresponding author: e-mail address: v.glover@imperial.ac.uk

Contents

Abstract

The impact of stress on brain health begins in the womb. Both animal and human studies have found that prenatal maternal stress affects the brain and behavior of the offspring. Stressful life events, exposure to a natural disaster, and symptoms of maternal anxiety and depression increase the risk for the child having a range of emotional, behavioral and/or cognitive problems in later life. These include depression, anxiety, Attention Deficit Hyperactivity Disorder (ADHD), and/or conduct disorders. There is an increased risk for other outcomes also, including preterm delivery and reduced telomere length, possibly indicative of an accelerated life history. The causal role of prenatal maternal stress on the etiology of the neurodevelopmental disorders is supported by large population cohorts, which have controlled for a wide range of potential confounders, including postnatal maternal mood. More recently, research has begun to explore the biological correlates and mediators of these findings. These studies suggest that the hypothalamic pituitary adrenal (HPA) axis plays a role in mediating the effects of maternal stress on the fetal brain. Further, in vivo brain imaging research reports that maternal stress is associated with changes in limbic and frontotemporal networks, and the functional and microstructural connections linking them.

International Review of Neurobiology, Volume 150
ISSN 0074-7742
https://doi.org/10.1016/bs.irn.2019.11.002

The structural changes include cortical thinning and an enlarged amygdala. While these studies have been conducted on smaller sample sizes and could not control for many confounders, the observed brain changes do plausibly underlie many of the emotional, behavioral and cognitive changes found to be associated with prenatal stress.

1. Introduction

Research from the last 20 years has reported that maternal stress during pregnancy is associated with an increased risk of emotional, behavioral and cognitive problems in the offspring. These include symptoms of anxiety and depression, ADHD, conduct disorders (O'Connor, Heron, Golding, Beveridge, & Glover, 2002; O'Donnell, Glover, Barker, & O'Connor, 2014), personality disorders (Winsper, Wolke, & Lereya, 2015), cognitive problems (Pearson et al., 2016), and autism spectrum disorders (Kinney, Miller, Crowley, Huang, & Gerber, 2008). Prenatal stress can also cause changes in several other aspects of development. These include an acceleration of life history, with an increased risk of preterm delivery (Wadhwa et al., 2001), earlier menarche in girls (Duchesne, Liu, Jones, Laplante, & King, 2017), and reduced telomere length suggestive of a shortened lifespan (Entringer, de Punder, Buss, & Wadhwa, 2018). Studies also report alterations in immune function (Hahn et al., 2019), increased risk of asthma (Cookson, Granell, Joinson, Ben-Shlomo, & Henderson, 2009), an altered pattern of the microbiome in the child (Hu et al., 2019), and an altered sex ratio at birth with more girls born than boys (Walsh et al., 2019).

These effects may depend on other factors including the type of stress, the time in gestation that the stress is experienced, and the sex of the fetus (Glover & Hill, 2012). For example, the risk of schizophrenia has been found to be especially associated with severe stress in the first trimester (Guo, He, Song, & Zheng, 2019). Interpretation of these findings has been complicated by the potential confounding influence of genetic and perinatal environmental factors, which also can affect child outcome. The latter, for example, includes exposure to prenatal smoking and/or alcohol, maternal socioeconomic status, education, and postnatal maternal and paternal mood. However, large population studies that have controlled for these putative confounding influences still suggest an additional direct effect of prenatal stress on child outcome (Glover, 2019; King & Laplante, 2005; O'Donnell et al., 2014). Not all children are affected by prenatal stress (O'Donnell et al., 2014), and those that are can be affected in different ways.

This is due in part to the specific genetic vulnerabilities of each child, and gene environment interactions. For example, O'Donnell et al. (2017) showed that the child's working memory at 8 years, and a diagnosis of ADHD at 15 years, were only affected by maternal prenatal anxiety if the child had a specific form of COMT, the enzyme which metabolizes catecholamines.

1.1 Fetal programming

The fetal programming theory proposes that the environment in the womb, during different sensitive periods for specific outcomes, can alter the development of the fetus, with a long lasting effect on the child (Barker, 2003). Barker and colleagues developed this concept in relation to their findings that low birthweight, or growing less well in the womb, is associated with vulnerability to cardiovascular disease and the metabolic syndrome later in life. But the concept also applies to the development of the brain. The brain is particularly plastic, or sensitive to environmental influences, in the fetal and early postnatal period. However, changes in the womb do not mean that they cannot be altered again later. For example, some of the neurodevelopmental effects of prenatal life event stress or raised in utero cortisol can be buffered by sensitive attachment between the mother and the child postnatally (Bergman, Sarkar, Glover, & O'Connor, 2010).

Fetal programming can be part of a "predictive adaptive response" (Gluckman, Hanson, & Spencer, 2005; Hill, Pickles, Wright, Braithwaite, & Sharp, 2019). These predictive responses may be "plastic decisions made by the embryo/fetus/neonate in response to how it interprets the current environment as a predictor of the future one" (Gluckman, Beedle, Buklijas, Low, & Hanson, 2016). The idea is that the developing fetus adapts for future advantage in a predicted later environment, rather than for immediate survival. Many of the fetal brain changes that are observed in response to prenatal stress, may be adaptive in causing changes in behavior that helped survival for our ancestors in a dangerous environment. For example, an increase in the function of the amygdala, may underlie an increase in anxiety which is associated with an increase in vigilance and a more rapid detection of danger. This could be adaptive and help survival in an environment of real external physical danger, such as in the presence of a predator. The accelerated life history may also be adaptive in conditions of external danger, allowing a greater chance of survival and reproduction.

1.2 The fetal brain

The fetal period is critical for the development of the human brain, and the way it develops can be influenced by the environment. The brain grows most rapidly during the fetal period and during the first year after birth; the newborn brain is about 36% of the volume of the adult brain (Pulli et al., 2019). Fetal brain development is divided into three phases: embryonic period (from conception to the eighth gestational week), early fetal period (up to mid-gestation), and late fetal period (from mid-gestation until birth) (Monk, Lugo-Candelas, & Trumpff, 2019). The developing brain contains billions of neurons, most of which are produced by mid-gestation. Throughout the fetal period, brain development involves neurone production, migration, connection, and differentiation. By 20 weeks, the end of the early fetal period, the brain has the necessary structures for mature functioning, although still with a smooth cortical plate. Neuronal migration peaks between gestational weeks 12 and 20. Myelin is first detected at between 20 and 28 weeks, and enhances the speed and accuracy of neuronal communication. During the later fetal period, the second half of gestation, there is specialization in the different brain regions, and connective pathways and synapses start to form. At week 34, approximately 40,000 new synapses are formed every second, a process that continues into early postnatal life (Pulli et al., 2019). The fetal brain is thus under construction throughout gestation, and its development can be affected by the biological signals it receives. The effects of external chemicals such as tobacco, nicotine, cocaine and certain other drugs are well known. Also important are the endogenous biological signals from the mother, including those caused by maternal obesity (Li et al., 2016; Pulli et al., 2019), immune system function, or alterations caused by her mood or stress. The time in gestation that the stress is experienced, and the stage of fetal brain development, may affect the consequences that it has.

2. Effects of prenatal stress on fetal brain development

2.1 Animal studies

Research with a variety of animals including rats (Maccari et al., 2003; Maccari, Krugers, Morley-Fletcher, Szyf, & Brunton, 2014), guinea pigs (Bennett et al., 2017), and monkeys (Schneider et al., 1998, 2017) has shown that inducing prenatal stress in the laboratory can have enduring effects on the neurodevelopment of the offspring, with alterations in both brain and behavior. Animal studies can control for postnatal effects by cross-fostering the offspring of prenatally stressed mothers to new mothers who were not prenatally stressed. Increased anxiety and impaired cognition have been frequently observed in the offspring, together with enhanced HPA responsiveness (Emack & Matthews, 2011). Sex differences are often reported too (Glover & Hill, 2012). In general, prenatal stress increases anxiety, depression and stress responses in female offspring rather than in male. Males are more likely to show learning and memory deficits. There is accumulating evidence that a range of epigenetic processes including histone modifications, DNA methylation, and small non-coding RNAs are involved (Hamada & Matthews, 2019).

Weinstock (2017) has reviewed studies of the effects of prenatal stress on neurodevelopment and behavior in rats and mice. Stressing pregnant rodents causes changes in offspring behavior analogous to those observed in humans, including increased anxiety and depression. Deficits in spatial learning and memory have also been shown. Lee, Kim, and Goto (2016) have explored some of the complexities in these effects, showing different outcomes depending both on the nature of the stressor and the nature of the postnatal environment. Pregnant mice were subjected to repeated social defeat stress or restraint stress, and the offspring tested for cognitive and affective functioning. Adult mice exposed to prenatal stress exhibited heightened anxiety, while spatial memory was impaired only by prenatal restraint stress, and not social defeat stress. Some of the neurodevelopmental changes associated with prenatal stress could be counteracted by the postnatal environment (Morley-Fletcher, Rea, Maccari, & Laviola, 2003).

Several of the animal studies of the effects of prenatal stress on the offspring have also studied brain structure (Petit et al., 2015; Stevens, Su, Yanagawa, & Vaccarino, 2013) and connectivity (Goelman, Ilinca, Zohar, & Weinstock, 2014) and found changes similar to those observed in humans. Changes in the developmental trajectory of limbic/paralimbic brain regions central to attention/cognitive processing and emotional/social behavior, including the hippocampus and amygdala have been described (Kraszpulski, Dickerson, & Salm, 2006; Monk et al., 2019). In a rat model, prenatal cold stress both inhibited the growth and development of hippocampal neurons, and induced anxiety-like behavior in the offspring (Lian et al., 2019). Hwang, Ku, and Hashimoto-Torii (2019) have reviewed a range of animal studies discussing how prenatal exposure to various types of stress, including alcohol, drugs, and inflammation, disrupts neuronal migration and causes neuronal migration disorders.

2.2 Human studies

Most of the human studies have used magnetic resonance imaging (MRI) to study the effects of prenatal stress on the brain of the child. However, a few have used electroencephalography (EEG) (Adamson, Letourneau, & Lebel, 2018). Maternal depression during pregnancy has been related to right frontal EEG asymmetry in neonates (Diego et al., 2004), which has been suggested to be an indicator of risk for developing negative affectivity and behavioral inhibition. Changes in right frontal EEG activation have also been reported in relation to anxiety symptoms (Field et al., 2003); these types of changes have been associated with withdrawal behavior.

Several types of MRI have been used to examine the effects of prenatal stress on human brain development, including structural, diffusion, and functional MRI. These study the basic structure, white matter connectivity and microstructure, and neuronal activity, respectively (Jenkinson & Chappell, 2018). Neuronal activity can be studied both at rest (resting state fMRI) and during a task (task fMRI). The age of the offspring in the MRI scans has varied considerably, including adolescents or adults (Favaro, Tenconi, Degortes, Manara, & Santonastaso, 2015; Marečková et al., 2018), young children (Buss, Davis, Muftuler, Head, & Sandman, 2010; Wen et al., 2017) and infants or newborn (Lautarescu et al., 2019; Rifkin-Graboi et al., 2013).

Pulli et al. (2019) have reviewed some of these MRI studies and comment that many possible confounders including infant sex, birthweight, and age, and maternal prenatal alcohol, smoking, medication, SES, ethnicity, age and BMI can all affect outcomes and are often not all controlled for. Most of the studies are quite small with well under 100 subjects, and so it was not possible to control for multiple confounders, as the large populations studies, with several thousand subjects, looking at psychometric outcomes (e.g., O'Donnell et al., 2014) have done.

However, these MRI studies do give interesting insights into the structural and functional brain changes associated with prenatal stress, and how these may underlie many of the emotional, behavioral and cognitive changes observed. The findings are generally consistent with the larger population behavioral studies. Increased volume of the amygdala could underlie increases in anxiety or vigilance, and reduced cortical gray matter is consistent with depression and with cognitive problems. With the studies on newborns, any differences were not due to the nature of the postnatal care. Several studies have observed differences between males and females, as in the animals studies.

2.2.1 Structural MRI studies

Structural MRI provides information on the gross anatomy of the brain and investigates how prenatal stress may affect characteristics including local gray matter density, cortical thickness, and the shape and size of anatomical structures. Table 1 summarizes these studies.

The most consistently replicated findings are those of changes in the frontal and temporal lobes, including cortical thinning (Davis et al., 2019; El Marroun et al., 2016; Lebel et al., 2016; Sandman, Buss, Head, & Davis, 2015) and reductions in gray matter volume (Buss et al., 2010; Favaro et al., 2015; Marečková et al., 2018). One study has reported increased gray matter volume (i.e., in the left caudal middle frontal area, El Marroun et al., 2016). Decreased volume of the cerebellum has also been reported (Buss et al., 2010). Findings in the parietal lobe are inconsistent, with two studies reporting decreased volumes (Buss et al., 2010; Marečková et al., 2018) in children and adults, while one study reported increased volume in several areas of the parietal lobe in adolescents (McQuaid, Darcey, Avalos, Fishbein, & VanMeter, 2019). However, the latter findings were part of a study which used a retrospective measure of PTSD which was administered over 10 years after birth, without controlling for postnatal mood.

Changes have also been observed in the limbic system, with increases in amygdala volume reported in both children (Acosta et al., 2019; Wen et al., 2017) and adults (Jones et al., 2019). However no differences were found in neonates exposed to maternal stress (Lugo-Candelas et al., 2018; Rifkin-Graboi et al., 2013). Decreased volume of the hippocampus has been reported in children exposed to maternal anxiety (Qiu et al., 2013); and decreased volume of the anterior cingulate cortex in adults whose mothers experienced stressful life events during pregnancy (Marečková et al., 2018).

Some of these studies also highlight the possible importance of timing in regards to exposure to stress, and the age at which the child is assessed. Buss et al. (2010) reported changes in gray matter volume when children were exposed to maternal stress at 19 weeks gestational age but not at 25 and 31 weeks, while Sandman et al. (2015) reported associations with cortical thinning which were strongest for exposure at 25 weeks. One longitudinal study reported no differences in hippocampal volume at the neonatal time point, but found that the prenatally stressed group showed reduced hippocampal volume at 6 months, suggesting slower growth in the postnatal period (Qiu et al., 2013).

Table 1 Structural MRI studies.

Outcome in offspring	Prenatal stress assessment	Sample	References
No difference in right *amygdala* volume	EPDS (26 weeks)	n = 157 (neonates)	Rifkin-Graboi et al. (2013)
Slower growth in *hippocampal* volume (no difference at birth but smaller volumes at 6 months)	STAI (26 weeks, 3 months)	n = 175 (neonates) n = 35 (6 months)	Qiu et al. (2013)
No difference between untreated depression and controls. SSRI exposed: Increased gray matter volume in right amygdala and right insula.	CESDS, SADS (19/39 weeks)	n = 98 (infants)	Lugo-Candelas et al. (2018)
Cortical thinning (primarily in right *frontal* lobes). Association at 25 weeks	CESDS (19/25/31 weeks)	n = 81 (age 7 years)	Sandman et al. (2015)
No difference for history of depression but no SSRI. SSRI exposed -more likely to have Chiari I malformations	Diagnosis of depression/SSRI	n = 33 SSRI (66 controls), n = 30 history (60 controls), (age 1–2 years)	Knickmeyer et al. (2014)
Cortical thinning in right inferior *frontal* and middle *temporal* regions	EPDS (2nd trimester)	n = 52 (age 2.6–5.1 years)	Lebel et al. (2016)
Greater left relative amygdala volume in girls compared to boys. Association in second trimester	PRAQ-2	n = 27 (age 4 years)	Acosta et al. (2019)
Increased volume in right *amygdala* in girls but not boys	EPDS (26 weeks)	n = 235 (age 4.5 years)	Wen et al. (2017)

Decreased GM in prefrontal cortex, premotor cortex, medial temporal lobe, lateral temporal cortex, postcentral gyrus, cerebellum. Association at 19 weeks	PAS (19,25,31 weeks)	n = 35 (age 6–9 years)	Buss et al. (2010)
Cortical thinning in superior *frontal* cortex in left hemisphere. Larger caudal middle frontal area in left hemisphere	BSI (20 weeks)	n = 654 (age 6–10 years)	El Marroun et al. (2016)
Cortical thinning in *frontal* and *temporal* regions	PSS (average)	n = 74 (age 7 years)	Davis et al. (2019)
Increased volume in *amygdala* in girls	Impact of event scale (quebec ice storm)	n = 68 (age 11 years)	Jones et al. (2019)
Increased GM in posterior *parietal* cortex, bilateral intraparietal sulcus, left superior parietal lobule, inferior parietal lobule	PTSD (retrospective)	n = 83 (age 11–14 years)	McQuaid et al. (2019)
Lower overall GM volume in mid-dorsolateral frontal cortex, anterior cingulate cortex and precuneus. *Hippocampal volume* no association with prenatal stress, but with postnatal anxiety	Stressful life events, (20 weeks)	n = 93 (age 23–24 years)	Marečková et al. (2018)
Decreased GM in left medial *temporal* lobe and *amygdala*	Interview	n = 35 (age 25 years), females	Favaro et al. (2015)

BSI, brief symptom inventory; CESDS, centre for epidemiological studies depression scale; EPDS, edinburgh postnatal depression scale; PAS, pregnancy anxiety scale; PRAQ, pregnancy related anxiety; SADS, schedule for affective disorders and schizophrenia; STAI, state trait anxiety inventory; GM, gray matter.

Adapted from Scheinost, D., Sinha, R., Cross, S. N., Kwon, S. H., Sze, G., Constable, R. T., et al. (2017). Does prenatal stress alter the developing connectome? *Pediatric Research, 81*(1–2), 214–226.

Most of this research did not account for relevant potential confounders such as postnatal maternal depression. However, some report that the associations were independent of postnatal stress (Buss et al., 2010; Favaro et al., 2015). Some also found different effects of prenatal stress depending on the sex of the child. Wen et al. (2017) and Acosta et al. (2019) reported larger amygdala volume in girls, but not boys, and Jones et al. (2019) found that when controlling for postnatal factors prenatal stress was associated with a larger amygdala volumes in girls, but not in boys. One study only examined females (Favaro et al., 2015)

Some studies have examined the effects of SSRIs. Babies exposed to SSRIs in utero showed increase gray matter volume in the limbic system (Lugo-Candelas et al., 2018), increased rates of Chiari I malformations (Knickmeyer et al., 2014), and changes in white matter microstructure (Jha et al., 2016) and connectivity (Lugo-Candelas et al., 2018). This highlights the importance of controlling for psychotropic medication in studies of maternal prenatal stress.

A few of these studies have associated the observed brain changes with the behavior of the offspring. Cortical thinning in prefrontal areas of the right hemisphere was associated with externalizing behavior at 7 years (Sandman et al., 2015) and cortical thinning, primarily in frontal and temporal regions, was associated with elevated depressive symptoms in adolescence (Davis et al., 2019). Cortical thinning has been found in major depressive disorder independently of prenatal stress (Suh et al., 2019). The amygdala is a brain structure involved in emotional regulation, and larger amygdala volumes are associated with behavioral disorders. Jones et al. (2019) reported that the subjective stress of being exposed to a Canadian Ice storm in late pregnancy was associated with a larger amygdala volume and this explained higher levels of externalizing behavior at 11 years. When adjusting for postnatal factors, the effect was only significant in girls. There were no associations with internalizing behavior. Acosta et al. (2019) reported that pregnancy-specific anxiety was associated with larger left amygdala volumes, which was linked to more emotional symptoms and overall difficulties.

2.2.2 Diffusion MRI studies

One of the mechanisms by which prenatal maternal stress may affect fetal brain development is by disruption of the white matter circuitry (Sarkar et al., 2014), and this can be shown by diffusion MRI. Diffusion MRI characterizes water molecular motion and is sensitive to the microstructural properties of tissue, such as axon size, density and myelination, as well as being able to provide information on anatomical connectivity, the wiring of the brain (Jenkinson & Chappell, 2018) White matter has a prolonged maturation which continues from the prenatal period until well after birth and into later life (Dubois et al., 2014). The biological significance of diffusion MRI findings is often unclear, but low anisotropy and high diffusivity are typically believed to reflect lower microstructural organization, reductions in myelination, and axonal damage. Table 2 summarizes the studies which have examined the relationship between maternal prenatal stress and diffusion metrics.

Despite some inconsistencies, the overall results suggest that prenatal stress is associated with white matter changes in limbic and frontal areas

Table 2 Diffusion MRI studies.

Outcome in offspring	Stress assessment	Sample	References
Increased MD, AD, RD in left *uncinate fasciculus* and increased AD in right uncinate fasciculus	STAI, stressful life events (prenatal)	n = 251 (premature neonates)	Lautarescu et al. (2019)
Decreased FA in *right amygdala*	EPDS (26 weeks)	n = 157 (neonates)	Rifkin-Graboi et al. (2013)
Decreased FA in a range of areas (insula, DLPFC, right middle occipital cortex, angular gyrus, uncinate, posterior cingulate, etc.)	STAI (26 weeks)	n = (54 neonates)	Rifkin-Graboi et al. (2015)
Decreased structural connectivity between right *amygdala* and right ventral *PFC*	CESDS (32 weeks)	n = 64 (neonates)	Posner et al. (2016)
No difference for untreated depression. For SSRI exposed: Lower FA, increased MD and RD for multiple fiber bundles	Diagnosis of depression/ SSRI use	n = 54 history, n = 27 SSRI, + matched controls (neonates)	Jha et al. 2016
Decreased neurite density, increased MD, RD, AD in *right frontal* white matter microstructure	EPDS, STAI (28 & 35 weeks)	n = 101 infants	Dean et al. (2018)

Continued

Table 2 Diffusion MRI studies.—cont'd

Outcome in offspring	Stress assessment	Sample	References
No difference between untreated depression and controls. For SSRI exposed: Increased connectivity between right amygdala and right insula	CESDS, SADS (19–39 weeks)	n =98 infants	Lugo-Candelas et al. (2018)
Decreased RD and MD in white matter emanating from inferior *frontal* area. Did not survive correction for postpartum EPDS	EPDS (11,17,32 weeks)	n=52 (age 2.6–5.1 years)	Lebel et al. (2016)
Decreased FA, Increased MD in *amygdala-frontal* tract and *cingulum*. Associations for thirrd trimester	EPDS (11,16,31 weeks)	n=54 (age 4 years)	Hay et al. (2019)
Higher right amygdala FA in overall sample and in girls but not boys	EPDS (26 weeks)	n= 235 (age 4.5 years)	Wen et al. (2017)
Increased FA and *decreased RD* in right *uncinate fasciculus*	Life events (prenatal)	n=22 (age 7 years)	Sarkar et al. (2014)
Decreased FA in left *cingulum*	Stressful life events (20 weeks)	n =93 (age 22–23 years)	Marečková et al. (2018)

CESDS, centre for epidemiological studies depression scale; EPDS, edinburgh postatal depression scale; STAI, state trait anxiety inventory; SLE, stressful life events; MD, mean diffusivity; AD, axial diffusivity; RD, radial diffusivity; FA, fractional anisotropy; SSRI, selective serotonin reuptake inhibitors; PFC, prefrontal cortex; DLPFC, dorsolateral prefrontal cortex.

Adapted from Scheinost, D., Sinha, R., Cross, S. N., Kwon, S. H., Sze, G., Constable, R. T., et al. (2017). Does prenatal stress alter the developing connectome? *Pediatric Research, 81*(1–2), 214–226.

in the child's brain. The studies have examined different forms of prenatal stress, and different tracts, and at different ages of the child. Increases in diffusivity (mean diffusivity, radial diffusivity, or axial diffusivity) have been reported in the amygdala (Wen et al., 2017), frontal white matter (Dean et al., 2018; Lebel et al., 2016), amygdala-frontal tracts (Hay et al., 2019) and the uncinate fasciculus (Lautarescu et al., 2019; Sarkar et al., 2014) as well as the cingulum (Hay et al., 2019, Marečková et al., 2018). Reductions in fractional anisotropy (FA) have been reported in a range of limbic and frontal areas (Hay et al., 2019; Rifkin-Graboi et al., 2015).

The finding of changes in the FA of the amygdala have differed depending on the age of the child. Prenatal stress exposure was associated with a decrease in FA in a sample of neonates (Rifkin-Graboi et al., 2013), but an increase in FA in a sample of children (Wen et al., 2017).

The uncinate fasciculus connects the amygdala with the frontal cortex and abnormalities in this tract in children and adults have been associated with antisocial behavior and mood disorders. It is thus of interest that this tract has been shown to be abnormal following prenatal stress in different studies, one in preterm neonates (Lautarescu et al., 2019) and one in 7-year-old children (Sarkar et al., 2014). The effects in this tract were compared in both of these studies with those in another tract, the inferior longitudinal fasciculus, which has not been consistently reported to be involved in social and emotional behavior, and which as predicted, showed no changes in diffusion MRI. Other studies have also reported prenatal stress-related changes in areas including the uncinate fasciculus, with reductions in FA (Rifkin-Graboi et al., 2013, 2015) and higher diffusivity (Dean et al., 2018).

A decrease in neurite density has been observed in the right frontal white matter of infants exposed to prenatal stress (Dean et al., 2018). Posner et al. (2016) found a decrease in structural connectivity between right amygdala and right ventral prefrontal cortex. Lebel et al. (2016) reported an inverse correlation between maternal prenatal EPDS scores and radial diffusivity and mean diffusivity in the white matter of inferior frontal areas in children, but these results lost significance when controlling for postnatal symptoms of depression.

As with the findings reported for structural MRI, sex appears to be an important factor, with maternal stress affecting the brain development of male and female offspring in different ways (Dean et al., 2018). Some studies have reported results observed only in females (Wen et al., 2017), or only in males (Hay et al., 2019).

Hay et al. (2019) have recently shown that white matter microstructure changes in the right amygdala pathway were associated with externalizing behavioral symptoms, and in males these changes in white matter development mediated the relationship between third trimester depressive symptoms and this externalizing behavior.

2.2.3 Functional MRI studies

Changes in the neurodevelopment of offspring exposed to prenatal stress can also be studied by using resting state fMRI (rs-fMRI) or task fMRI. fMRI infers dynamic changes in brain activity by measuring hemodynamic changes (i.e., BOLD, blood oxygenation level dependent effect) (Jenkinson & Chappell, 2018). Rs-fMRI provides information about spontaneous brain activity and the functional connectivity of brain regions, even if they are not directly linked structurally (Scheinost et al., 2017). Task-related fMRI analyses allow researchers to locate and analyse brain activity in response to a task. Studies of prenatal stress and functional brain connectivity have, as with diffusion MRI, primarily found changes in the associations between the amygdala and other regions (Table 3).

Qiu et al. (2015) have found increased connectivity between the amygdala and several areas including the left temporal cortex, insula, and medial orbitofrontal and ventromedial prefrontal cortices in 6 month old babies exposed to prenatal symptoms of depression. Scheinost et al. (2016) have

Table 3 Resting state functional MRI studies.

Outcome in offspring	Stress assessment	Sample	References
Reduced connectivity between left amygdala and subcortical regions (thalamus, hypothalamus and peristriate cortex)	Retrospective review of chart	n=26 (preterm neonates)	Scheinost et al. (2016)
Increased negative connectivity between amygdala (right and left) and dorsal prefrontal cortex.	CESDS (32 weeks)	n=64 (neonates)	Posner et al. (2016)
Increased connectivity between amygdala and several areas (left temporal cortex, insula, bilateral ACC, medial orbitofrontal ventromedial prefrontal cortex)	EPDS (26 weeks)	n=24 (age 6 months)	Qiu et al. (2015)
Reduced connectivity between left amygdala and bilateral subgenual ACC and left caudate, and between right amygdala and left OFC, insula and temporal pole. In girls but not in boys	EPDS (26 weeks)	n =128 (age 4.5 years)	Soe et al. (2018)
Increased connectivity between left medial temporal lobe and pregenual cortex	Interview	n=35 (age 25 years), females	Favaro et al. (2015)

CESDS, center for epidemiological studies depression scale; EPDS, edinburgh postnatal depression scale; OFC, orbitofrontal cortex; ACC, anterior cingulate cortex.

Adapted from Scheinost, D., Sinha, R., Cross, S. N., Kwon, S. H., Sze, G., Constable, R. T., et al. (2017). Does prenatal stress alter the developing connectome? *Pediatric Research, 81*(1–2), 214–226.

reported reduced connectivity between the amygdala and subcortical areas including the thalamus in a sample of preterm neonates and Posner et al. (2016) reported reduced connectivity between the amygdala and the dorsal prefrontal cortex in neonates exposed to prenatal depression. In a study of adults exposed to prenatal stress in utero (Favaro et al., 2015) increased connectivity was observed between the left medial temporal lobe and the pregenual cortex. Soe et al. (2018) found altered functional connectivity between the amygdala and cortico-striatal circuitry in young children after exposure to prenatal depression in girls but not in boys, again emphasizing potential sex differences in the effects of prenatal stress. This pathway is essential for emotional perception and regulation.

There have been fewer task functional MRI studies. Children born to mothers with prenatal depression exhibited increased amygdala responses when presented with pictures of negative emotional faces compared with controls (van der Knaap et al., 2018). Sarkar and colleagues (submitted for publication) have investigated brain networks in 7-year-old children exposed to varying degrees of prenatal life event stress, using a rewarded Continuous Performance Task. They found associations between prenatal stress with later aberrations in brain function in a network of regions crucial for sustained attention and reward. Mennes, Stiers, Lagae, and Van den Bergh (2019) used fMRI to examine activity in response to a task indicative of cognitive control in 20-year-old men. They found an association between prenatal maternal anxiety and brain activity in a number of prefrontal cortical areas, in particular a number of right lateralized clusters. They describe endogenous cognitive control as the ability to generate control over decisions, strategies, conflicting information and so on, from within oneself, without external signals, and suggest that the differences in brain activity that they observed may underlie problems with such cognitive control.

2.3 Different types of stress

Tables 1–3 show that many different types of prenatal stress have been found to be associated with alterations in the brain development of the child. These range from symptoms of anxiety (Mennes et al., 2019; Qiu et al., 2013)' pregnancy-specific anxiety (Acosta et al., 2019; Buss et al., 2010), symptoms of depression (Ong et al., 2019), life events (Sarkar et al., 2014), or the natural disaster of a Canadian ice storm (Jones et al., 2019).

This is similar to the findings with prenatal stress on the offspring's psychopathology, which also have been shown to be altered by many types of stress. An increased risk of schizophrenia is unusual as it is only associated with very severe stress. It has been associated with exposure to an earthquake in China (Guo et al., 2019), the death of a close relative (Khashan et al., 2008), or the Dutch Hunger Winter (Susser et al., 1996), but only with exposure in the first trimester. In schizophrenia there is an altered migration pattern of the neurons, and this migration largely takes place in early to mid-gestation (Monk et al., 2019).

The rest of the brain continues to develop connections all through pregnancy and beyond. Many types of stress can occur together, such as symptoms of anxiety and depression. In some behavioral or cognitive studies where both have been measured together, symptoms of anxiety were associated with more adverse outcomes than those of depression (e.g., O'Donnell et al., 2017). There is some evidence that pregnancy-specific anxiety (worry about the outcome of the pregnancy for the mother or the child) is a distinct syndrome (Huizink, Mulder, Robles de Medina, Visser, & Buitelaar, 2004), and has some effects independently of other forms of anxiety or stress. These include altered methylation pattern of the glucocorticoid receptor in the newborn cord blood (Hompes et al., 2013), and the pattern of the microbiome in the newborn meconium (Hu et al., 2019). It has also been found to be associated with temperamental variation in young infants (Huizink, de Medina, Mulder, Visser, & Buitelaar, 2002) and their cognitive development (Huizink, Robles de Medina, Mulder, Visser, & Buitelaar, 2003). Buss et al. (2010) have shown that high levels of maternal pregnancy-specific anxiety were associated with lower inhibitory control in girls and lower visuospatial working memory performance in both boys and girls. They found that neither state anxiety nor depression explained any additional variance after accounting for pregnancy-specific anxiety (Buss, Davis, Hobel, & Sandman, 2011). No studies have so far determined the cause of this specific effect of pregnancy-specific anxiety, independently of other symptoms of anxiety. It would be interesting to investigate whether it was associated with previous experiences of the mother, including early trauma or abuse.

In general, it is not known whether different types of stress have any different effects on early brain development. It is possible that this may underlie some of the inconsistencies in the effects of prenatal stress in the literature, but this is still unknown.

3. Underlying mechanisms

Some of the biological mechanisms for the effects of prenatal stress on the later behavior and cognition of the offspring are starting to be understood (Monk et al., 2019). Based on work on animal models, many of the studies of underlying mechanisms in humans have focused on the HPA axis and the role of cortisol. In rodent models, removing the adrenal gland or injecting corticosteroids in the pregnant dam can block or replicate some of the effects of prenatal stress on the offspring (Zagron & Weinstock, 2006). This is the most studied system in humans also, although many other pathways are likely to be involved.

Maternal exposure to synthetic glucocorticoids during pregnancy is associated with poorer mental health in childhood and adolescence (Khalife et al., 2013). Bergman et al. (2010) showed that altered cognitive function in infants was associated with higher levels of cortisol in the amniotic fluid. The children of mothers who had consumed high levels of liquorice during pregnancy (which contains a natural inhibitor of 11-β-hydroxysteroid dehydrogenase type II (11β-HSD2), the enzyme which converts cortisol to its inactive form cortisone in the placenta), and were thus exposed to higher levels of cortisol in utero, were over three times more likely to show ADHD symptoms, as well as lower IQ scores and earlier puberty (Raikkonen et al., 2017).

The function of the HPA axis can be changed by many different types of stress, and this may explain why such varied stressors can be involved in the fetal programming of the brain. However, it is not known which stress-induced biological changes in the mother are involved. Maternal cortisol responses may be part of the mechanism, but not the major part. Unlike in most species, in humans, the placenta releases large amounts of CRH into the maternal circulation in all mothers as pregnancy advances, and this in turn releases cortisol. This makes the mother's HPA axis less sensitive to external stress as gestation increases (Bleker, Roseboom, Vrijkotte, Reynolds, & de Rooij, 2017). Her immune system and proinflammatory cytokines may well be involved but more evidence is needed for this. A history of early maternal trauma has been associated with elevated levels (Blackmore et al., 2011).

However, the fetus may be exposed to raised levels of cortisol independently of changes in maternal levels, when the mother is feeling stressed, because of increased passage of cortisol through the placenta.

Glover, Bergman, Sarkar, and O'Connor (2009) have found a greater correlation between maternal and amniotic fluid levels of cortisol when the mother had higher levels of anxiety. The fetal levels will thus also be higher if the maternal levels are. Several research groups have found that prenatal stress or anxiety is associated with an altered expression of several enzymes in the placenta, including a down regulation of 11 b-HSD2, the enzyme which metabolizes cortisol (e.g., O'Donnell et al., 2012).

Only a few studies have associated maternal or amniotic fluid cortisol levels with the brain structure of the child. Buss et al. (2012) have looked at associations with maternal prenatal cortisol levels. After accounting for the effects of some potential confounding pre- and postnatal factors, higher maternal blood cortisol levels in earlier but not later gestation were found to be associated with a larger right amygdala volume in 7-year-old girls but not in boys. Also, higher maternal cortisol levels in early gestation were associated with more affective problems in the girls, and this was mediated, in part, by the amygdala volume. Graham et al., 2019 have found that higher maternal cortisol levels were associated with stronger amygdala connectivity to brain regions involved in sensory processing and integration, as well as the default mode network in girls, and with weaker connectivity to these brain regions in boys. Sandman et al. (2018) examined the association between maternal blood CRH levels during pregnancy and the brain structure of the child at 7 years. They found an association between CRH levels throughout pregnancy and cortical thinning, primarily in the temporal and frontal regions. These associations were also stronger in girls. The authors were not able to show a causal link between the CRH levels and the cortical thinning, but propose it as a possibility. Sarkar and colleagues (submitted for publication) probed brain networks in 7-year-old children exposed to varying degrees of prenatal life event stress, using a rewarded Continuous Performance Task. They also had measures of amniotic fluid cortisol. The associations observed with the cortisol were overlapping but different from those with the life event stress, suggesting that exposure to higher levels of cortisol in utero may explain some, but not all, of the brain changes.

Even though most of the research on the underlying biological mechanisms for the effects of prenatal stress on fetal brain development has been on the HPA axis, the immune system is very likely to be involved also. Rasmussen et al. (2019) have shown that maternal interleukin-6 concentration during pregnancy was associated with variations in the fronto-limbic circuitry in the newborn, and this in turn correlated with the infant's cognition at 12 months.

4. Conclusions

There is considerable evidence that prenatal stress plays a causal role in a range of neurodevelopmental disorders. The brain changes underlying this are starting to be uncovered although many questions remain. Do different types of stress have different effects? To what extent are boys and girls affected differently? And how can prenatal effects on the brain be altered postnatally?

References

Acosta, H., Tuulari, J. J., Scheinin, N. M., Hashempour, N., Rajasilta, O., Lavonius, T. I., et al. (2019). Maternal pregnancy-related anxiety is associated with sexually dimorphic alterations in amygdala volume in four-year-old children. *Frontiers in Behavioral Neuroscience*, *13*, 175.

Adamson, B., Letourneau, N., & Lebel, C. (2018). Prenatal maternal anxiety and children's brain structure and function: A systematic review of neuroimaging studies. *Journal of Affective Disorders*, *241*, 117–126.

Barker, D. J. (2003). The developmental origins of adult disease. *European Journal of Epidemiology*, *18*(8), 733–736.

Bennett, G. A., Palliser, H. K., Shaw, J. C., Palazzi, K. L., Walker, D. W., & Hirst, J. J. (2017). Maternal stress in pregnancy affects myelination and neurosteroid regulatory pathways in the guinea pig cerebellum. *Stress*, *20*(6), 580–588.

Bergman, K., Sarkar, P., Glover, V., & O'Connor, T. G. (2010). Maternal prenatal cortisol and infant cognitive development: Moderation by infant-mother attachment. *Biological Psychiatry*, *67*(11), 1026–1032.

Blackmore, E. R., Moynihan, J. A., Rubinow, D. R., Pressman, E. K., Gilchrist, M., & O'Connor, T. G. (2011). Psychiatric symptoms and proinflammatory cytokines in pregnancy. *Psychosomatic Medicine*, *73*(8), 656–663.

Bleker, L. S., Roseboom, T. J., Vrijkotte, T. G., Reynolds, R. M., & de Rooij, S. R. (2017). Determinants of cortisol during pregnancy—The ABCD cohort. *Psychoneuroendocrinology*, *83*, 172–181.

Buss, C., Davis, E. P., Hobel, C. J., & Sandman, C. A. (2011). Maternal pregnancy-specific anxiety is associated with child executive function at 6–9 years age. *Stress*, *14*(6), 665–676.

Buss, C., Davis, E. P., Muftuler, L. T., Head, K., & Sandman, C. A. (2010). High pregnancy anxiety during mid-gestation is associated with decreased gray matter density in 6–9-year-old children. *Psychoneuroendocrinology*, *35*(1), 141–153.

Buss, C., Davis, E. P., Shahbaba, B., Pruessner, J. C., Head, K., & Sandman, C. A. (2012). Maternal cortisol over the course of pregnancy and subsequent child amygdala and hippocampus volumes and affective problems. *Proceedings of the National Academy of Sciences of the United States of America*, *109*(20), E1312–E1319.

Cookson, H., Granell, R., Joinson, C., Ben-Shlomo, Y., & Henderson, A. J. (2009). Mothers' anxiety during pregnancy is associated with asthma in their children. *The Journal of Allergy and Clinical Immunology*, *123*(4), 847–853, e811.

Davis, E. P., Hankin, B. L., Glynn, L. M., Head, K., Kim, D. J., & Sandman, C. A. (2019). Prenatal maternal stress, child cortical thickness, and adolescent depressive symptoms. *Child Development*, in press. https://doi.org/10.1111/cdev.13252.

Dean, D. C., Planalp, E. M., Wooten, W., Kecskemeti, S. R., Adluru, N., Schmidt, C. K., et al. (2018). Association of prenatal maternal depression and anxiety symptoms with infant white matter microstructure. *JAMA Pediatrics*, *172*(10), 973–981.
Diego, M. A., Field, T., Hernandez-Reif, M., Cullen, C., Schanberg, S., & Kuhn, C. (2004). Prepartum, postpartum, and chronic depression effects on newborns. *Psychiatry*, *67*(1), 63–80.
Dubois, J., Dehaene-Lambertz, G., Kulikova, S., Poupon, C., Huppi, P. S., & Hertz-Pannier, L. (2014). The early development of brain white matter: A review of imaging studies in fetuses, newborns and infants. *Neuroscience*, *276*, 48–71.
Duchesne, A., Liu, A., Jones, S. L., Laplante, D. P., & King, S. (2017). Childhood body mass index at 5.5 years mediates the effect of prenatal maternal stress on daughters' age at menarche: Project ice storm—ERRATUM. *Journal of Developmental Origins of Health and Disease*, *8*(3), 382.
El Marroun, H., Tiemeier, H., Muetzel, R. L., Thijssen, S., van der Knaap, N. J., Jaddoe, V. W., et al. (2016). Prenatal exposure to maternal and paternal depressive symptoms and brain morphology: A population-based prospective neuroimaging study in young children. *Depression and Anxiety*, *33*(7), 658–666.
Emack, J., & Matthews, S. G. (2011). Effects of chronic maternal stress on hypothalamo–pituitary–adrenal (HPA) function and behavior: No reversal by environmental enrichment. *Hormones and Behavior*, *60*(5), 589–598.
Entringer, S., de Punder, K., Buss, C., & Wadhwa, P. D. (2018). The fetal programming of telomere biology hypothesis: An update. *Philosophical Transactions of the Royal Society of London. Series B, Biological Sciences*, *373*, 20170151.
Favaro, A., Tenconi, E., Degortes, D., Manara, R., & Santonastaso, P. (2015). Neural correlates of prenatal stress in young women. *Psychological Medicine*, *45*(12), 2533–2543.
Field, T., Diego, M., Hernandez-Reif, M., Schanberg, S., Kuhn, C., Yando, R., et al. (2003). Pregnancy anxiety and comorbid depression and anger: Effects on the fetus and neonate. *Depression and Anxiety*, *17*(3), 140–151.
Glover, V. (2019). Is the association between maternal stress during pregnancy and the child's depression partly causal, and what should we do about it? *Acta Psychiatrica Scandinavica*, *139*(4), 301–303.
Glover, V., Bergman, K., Sarkar, P., & O'Connor, T. G. (2009). Association between maternal and amniotic fluid cortisol is moderated by maternal anxiety. *Psychoneuroendocrinology*, *34*(3), 430–435.
Glover, V., & Hill, J. (2012). Sex differences in the programming effects of prenatal stress on psychopathology and stress responses: An evolutionary perspective. *Physiology & Behavior*, *106*(5), 736–740.
Gluckman, P., Beedle, A., Buklijas, T., Low, F., & Hanson, M. (2016). *Principles of Evolutionary Medicine* (2nd ed.). Oxford University Press.
Gluckman, P. D., Hanson, M. A., & Spencer, H. G. (2005). Predictive adaptive responses and human evolution. *Trends in Ecology & Evolution*, *20*(10), 527–533.
Graham, A. M., Rasmussen, J. M., Entringer, S., Ben Ward, E., Rudolph, M. D., Gilmore, J. H., et al. (2019). Maternal cortisol concentrations during pregnancy and sex-specific associations with neonatal amygdala connectivity and emerging internalizing behaviors. *Biological Psychiatry*, *85*(2), 172–181.
Goelman, G., Ilinca, R., Zohar, I., & Weinstock, M. (2014). Functional connectivity in prenatally stressed rats with and without maternal treatment with ladostigil, a brain-selective monoamine oxidase inhibitor. *European Journal of Neuroscience*, *40*(5), 2734–2743.
Guo, C., He, P., Song, X., & Zheng, X. (2019). Long-term effects of prenatal exposure to earthquake on adult schizophrenia. *The British Journal of Psychiatry*, *215*(6), 1–6.
Hahn, J., Gold, D. R., Coull, B. A., McCormick, M. C., Finn, P. W., Perkins, D. L., et al. (2019). Prenatal maternal depression and neonatal immune responses. *Psychosomatic Medicine*, *81*(4), 320–327.

Hamada, H., & Matthews, S. G. (2019). Prenatal programming of stress responsiveness and behaviours: Progress and perspectives. *Journal of Neuroendocrinology*, *31*(3), e12674.

Hay, R. E., Reynolds, J. E., Grohs, M., Paniukov, D., Giesbrecht, G. F., Letourneau, N., et al. (2019). Examining the relationship between prenatal depression, amygdala-prefrontal structural connectivity and behaviour in preschool children. *bioRxiv*, 692335.

Hill, J., Pickles, A., Wright, N., Braithwaite, E., & Sharp, H. (2019). Predictions of children's emotionality from evolutionary and epigenetic hypotheses. *Scientific Reports*, *9*(1), 2519.

Hompes, T., Izzi, B., Gellens, E., Morreels, M., Fieuws, S., Pexsters, A., et al. (2013). Investigating the influence of maternal cortisol and emotional state during pregnancy on the DNA methylation status of the glucocorticoid receptor gene (NR3C1) promoter region in cord blood. *Journal of Psychiatric Research*, *47*(7), 880–891.

Hu, J., Ly, J., Zhang, W., Huang, Y., Glover, V., Peter, I., et al. (2019). Microbiota of newborn meconium is associated with maternal anxiety experienced during pregnancy. *Developmental Psychobiology*, *61*(5), 640–649.

Huizink, A. C., de Medina, P. G., Mulder, E. J., Visser, G. H., & Buitelaar, J. K. (2002). Psychological measures of prenatal stress as predictors of infant temperament. *Journal of the American Academy of Child and Adolescent Psychiatry*, *41*(9), 1078–1085.

Huizink, A. C., Mulder, E. J., Robles de Medina, P. G., Visser, G. H., & Buitelaar, J. K. (2004). Is pregnancy anxiety a distinctive syndrome? *Early Human Development*, *79*(2), 81–91.

Huizink, A. C., Robles de Medina, P. G., Mulder, E. J., Visser, G. H., & Buitelaar, J. K. (2003). Stress during pregnancy is associated with developmental outcome in infancy. *Journal of Child Psychology and Psychiatry*, *44*(6), 810–818.

Hwang, H. M., Ku, R. Y., & Hashimoto-Torii, K. (2019). Prenatal environment that affects neuronal migration. *Frontiers in Cell and Development Biology*, 7, 138.

Jenkinson, M., & Chappell, M. (2018). *Introduction to neuroimaging analysis*. Oxford University Press.

Jha, S. C., Meltzer-Brody, S., Steiner, R. J., Cornea, E., Woolson, S., Ahn, M., et al. (2016). Antenatal depression, treatment with selective serotonin reuptake inhibitors, and neonatal brain structure: a propensity-matched cohort study. *Psychiatry Research: Neuroimaging*, *253*, 43–53.

Jones, S. L., Dufoix, R., Laplante, D. P., Elgbeili, G., Patel, R., Chakravarty, M. M., et al. (2019). Larger amygdala volume mediates the association between prenatal maternal stress and higher levels of externalizing behaviors: Sex specific effects in project ice storm. *Frontiers in Human Neuroscience*, *13*, 144.

Khalife, N., Glover, V., Taanila, A., Ebeling, H., Jarvelin, M. R., & Rodriguez, A. (2013). Prenatal glucocorticoid treatment and later mental health in children and adolescents. *PLoS One*, *8*(11), e81394.

Khashan, A. S., Abel, K. M., McNamee, R., Pedersen, M. G., Webb, R. T., Baker, P. N., et al. (2008). Higher risk of offspring schizophrenia following antenatal maternal exposure to severe adverse life events. *Archives of General Psychiatry*, *65*(2), 146–152.

King, S., & Laplante, D. P. (2005). The effects of prenatal maternal stress on children's cognitive development: project ice storm. *Stress*, *8*(1), 35–45.

Kinney, D. K., Miller, A. M., Crowley, D. J., Huang, E., & Gerber, E. (2008). Autism prevalence following prenatal exposure to hurricanes and tropical storms in Louisiana. *Journal of Autism and Developmental Disorders*, *38*(3), 481–488.

Knickmeyer, R. C., Meltzer-Brody, S., Woolson, S., Hamer, R. M., Smith, J. K., Lury, K., et al. (2014). Rate of chiari I malformation in children of mothers with depression with and without prenatal SSRI exposure. *Neuropsychopharmacology*, *39*(11), 2611.

Kraszpulski, M., Dickerson, P. A., & Salm, A. K. (2006). Prenatal stress affects the developmental trajectory of the rat amygdala. *Stress*, *9*(2), 85–95.

Lautarescu, A., Pecheva, D., Nosarti, C., Nihouarn, J., Zhang, H., Victor, S., et al. (2019). Maternal prenatal stress is associated with altered uncinate fasciculus microstructure in premature neonates. *Biological Psychiatry*, [Epub ahead of print]. https://doi.org/10.1016/j.biopsych.2019.08.010.

Lebel, C., Walton, M., Letourneau, N., Giesbrecht, G. F., Kaplan, B. J., & Dewey, D. (2016). Prepartum and postpartum maternal depressive symptoms are related to children's brain structure in preschool. *Biological Psychiatry*, *80*(11), 859–868.

Lee, Y. A., Kim, Y. J., & Goto, Y. (2016). Cognitive and affective alterations by prenatal and postnatal stress interaction. *Physiology & Behavior*, *165*, 146–153.

Li, Y. M., Ou, J. J., Liu, L., Zhang, D., Zhao, J. P., & Tang, S. Y. (2016). Association between maternal obesity and autism spectrum disorder in offspring: A meta-analysis. *Journal of Autism and Developmental Disorders*, *46*(1), 95–102.

Lian, S., Xu, B., Wang, D., Wang, L., Li, W., Yao, R., et al. (2019). Possible mechanisms of prenatal cold stress induced-anxiety-like behavior depression in offspring rats. *Behavioural Brain Research*, *359*, 304–311.

Lugo-Candelas, C., Cha, J., Hong, S., Bastidas, V., Weissman, M., Fifer, W. P., et al. (2018). Associations between brain structure and connectivity in infants and exposure to selective serotonin reuptake inhibitors during pregnancy. *JAMA Pediatrics*, *172*(6), 525–533.

Maccari, S., Darnaudery, M., Morley-Fletcher, S., Zuena, A. R., Cinque, C., & Van Reeth, O. (2003). Prenatal stress and long-term consequences: Implications of glucocorticoid hormones. *Neuroscience and Biobehavioral Reviews*, *27*(1-2), 119–127.

Maccari, S., Krugers, H. J., Morley-Fletcher, S., Szyf, M., & Brunton, P. J. (2014). The consequences of early-life adversity: Neurobiological, behavioural and epigenetic adaptations. *Journal of Neuroendocrinology*, *26*(10), 707–723.

Marečková, K., Klasnja, A., Bencurova, P., Andrýsková, L., Brázdil, M., & Paus, T. (2018). Prenatal stress, mood, and gray matter volume in young adulthood. *Cerebral Cortex*, *29*(3), 1244–1250.

McQuaid, G. A., Darcey, V. L., Avalos, M. F., Fishbein, D. H., & VanMeter, J. W. (2019). Altered cortical structure and psychiatric symptom risk in adolescents exposed to maternal stress in utero: A retrospective investigation. *Behavioural Brain Research*, *375*, 112145.

Mennes, M., Stiers, P., Lagae, L., & Van den Bergh, B. R. H. (2019). Antenatal maternal anxiety modulates the BOLD response in 20-year-old men during endogenous cognitive control. *Brain Imaging and Behavior*, [Epub ahead of print]. https://doi.org/10.1007/s11682-018-0027-6.

Monk, C., Lugo-Candelas, C., & Trumpff, C. (2019). Prenatal developmental origins of future psychopathology: Mechanisms and pathways. *Annual Review of Clinical Psychology*, *15*, 317–344. [Epub ahead of print].

Morley-Fletcher, S., Rea, M., Maccari, S., & Laviola, G. (2003). Environmental enrichment during adolescence reverses the effects of prenatal stress on play behaviour and HPA axis reactivity in rats. *The European Journal of Neuroscience*, *18*(12), 3367–3374.

O'Connor, T. G., Heron, J., Golding, J., Beveridge, M., & Glover, V. (2002). Maternal antenatal anxiety and children's behavioural/emotional problems at 4 years. Report from the Avon Longitudinal Study of Parents and Children. *The British Journal of Psychiatry*, *180*, 502–508.

O'Donnell, K. J., Bugge Jensen, A., Freeman, L., Khalife, N., O'Connor, T. G., & Glover, V. (2012). Maternal prenatal anxiety and downregulation of placental 11beta-HSD2. *Psychoneuroendocrinology*, *37*(6), 818–826.

O'Donnell, K. J., Glover, V., Barker, E. D., & O'Connor, T. G. (2014). The persisting effect of maternal mood in pregnancy on childhood psychopathology. *Development and Psychopathology*, *26*(2), 393–403.

O'Donnell, K. J., Glover, V., Lahti, J., Lahti, M., Edgar, R. D., Raikkonen, K., et al. (2017). Maternal prenatal anxiety and child COMT genotype predict working memory and symptoms of ADHD. *PLoS One*, *12*(6), e0177506.

Ong, M. L., Tuan, T. A., Poh, J., Teh, A. L., Chen, L., Pan, H., et al. (2019). Neonatal amygdalae and hippocampi are influenced by genotype and prenatal environment, and reflected in the neonatal DNA methylome. *Genes, Brain, and Behavior, 18*(7), e12576.

Pearson, R. M., Bornstein, M. H., Cordero, M., Scerif, G., Mahedy, L., Evans, J., et al. (2016). Maternal perinatal mental health and offspring academic achievement at age 16: The mediating role of childhood executive function. *Journal of Child Psychology and Psychiatry, 57*(4), 491–501.

Petit, B., Boissy, A., Zanella, A., Chaillou, E., Andanson, S., Bes, S., et al. (2015). Stress during pregnancy alters dendritic spine density and gene expression in the brain of new-born lambs. *Behavioural Brain Research, 291*, 155–163.

Posner, J., Cha, J., Roy, A. K., Peterson, B. S., Bansal, R., Gustafsson, H. C., et al. (2016). Alterations in amygdala–prefrontal circuits in infants exposed to prenatal maternal depression. *Translational Psychiatry, 6*(11), e935.

Pulli, E. P., Kumpulainen, V., Kasurinen, J. H., Korja, R., Merisaari, H., Karlsson, L., et al. (2019). Prenatal exposures and infant brain: Review of magnetic resonance imaging studies and a population description analysis. *Human Brain Mapping, 40*(6), 1987–2000.

Qiu, A., Anh, T. T., Li, Y., Chen, H., Rifkin-Graboi, A., Broekman, B. F., et al. (2015). Prenatal maternal depression alters amygdala functional connectivity in 6-month-old infants. *Translational Psychiatry, 5*, e508.

Qiu, A., Rifkin-Graboi, A., Chen, H., Chong, Y. S., Kwek, K., Gluckman, P. D., et al. (2013). Maternal anxiety and infants' hippocampal development: Timing matters. *Translational Psychiatry, 3*, e306.

Raikkonen, K., Martikainen, S., Pesonen, A. K., Lahti, J., Heinonen, K., Pyhala, R., et al. (2017). Maternal licorice consumption during pregnancy and pubertal, cognitive, and psychiatric outcomes in children. *American Journal of Epidemiology, 185*(5), 1–12.

Rasmussen, J. M., Graham, A. M., Entringer, S., Gilmore, J. H., Styner, M., Fair, D. A., et al. (2019). Maternal Interleukin-6 concentration during pregnancy is associated with variation in frontolimbic white matter and cognitive development in early life. *NeuroImage, 185*, 825–835.

Rifkin-Graboi, A., Bai, J., Chen, H., Hameed, W. B., Sim, L. W., Tint, M. T., et al. (2013). Prenatal maternal depression associates with microstructure of right amygdala in neonates at birth. *Biological Psychiatry, 74*(11), 837–844.

Rifkin-Graboi, A., Meaney, M. J., Chen, H., Bai, J., Hameed, W. B., Tint, M. T., et al. (2015). Antenatal maternal anxiety predicts variations in neural structures implicated in anxiety disorders in newborns. *Journal of the American Academy of Child and Adolescent Psychiatry, 54*(4), 313–321.e2.

Sandman, C. A., Buss, C., Head, K., & Davis, E. P. (2015). Fetal exposure to maternal depressive symptoms is associated with cortical thickness in late childhood. *Biological Psychiatry, 77*(4), 324–334.

Sandman, C. A., Curran, M. M., Davis, E. P., Glynn, L. M., Head, K., & Baram, T. Z. (2018). Cortical thinning and neuropsychiatric outcomes in children exposed to prenatal adversity: A role for placental CRH? *American Journal of Psychiatry, 175*(5), 471–479.

Sarkar, S., Craig, M. C., Dell'Acqua, F., O'Connor, T. G., Catani, M., Deeley, Q., et al. (2014). Prenatal stress and limbic-prefrontal white matter microstructure in children aged 6–9 years: A preliminary diffusion tensor imaging study. *The World Journal of Biological Psychiatry, 15*(4), 346–352.

Scheinost, D., Kwon, S. H., Lacadie, C., Sze, G., Sinha, R., Constable, R. T., et al. (2016). Prenatal stress alters amygdala functional connectivity in preterm neonates. *NeuroImage. Clinical, 12*, 381–388.

Scheinost, D., Sinha, R., Cross, S. N., Kwon, S. H., Sze, G., Constable, R. T., et al. (2017). Does prenatal stress alter the developing connectome? *Pediatric Research, 81*(1–2), 214–226.

Schneider, M. L., Clarke, A. S., Kraemer, G. W., Roughton, E. C., Lubach, G. R., Rimm-Kaufman, S., et al. (1998). Prenatal stress alters brain biogenic amine levels in primates. *Development and Psychopathology*, *10*(3), 427–440.
Schneider, M. L., Moore, C. F., Adkins, M., Barr, C. S., Larson, J. A., Resch, L. M., et al. (2017). Sensory processing in rhesus monkeys: Developmental continuity, prenatal treatment, and genetic influences. *Child Development*, *88*(1), 183–197.
Soe, N. N., Wen, D. J., Poh, J. S., Chong, Y. S., Broekman, B. F., Chen, H., et al. (2018). Perinatal maternal depressive symptoms alter amygdala functional connectivity in girls. *Human Brain Mapping*, *39*(2), 680–690.
Stevens, H. E., Su, T., Yanagawa, Y., & Vaccarino, F. M. (2013). Prenatal stress delays inhibitory neuron progenitor migration in the developing neocortex. *Psychoneuroendocrinology*, *38*(4), 509–521.
Suh, J. S., Schneider, M. A., Minuzzi, L., MacQueen, G. M., Strother, S. C., Kennedy, S. H., et al. (2019). Cortical thickness in major depressive disorder: A systematic review and meta-analysis. *Progress in Neuro-Psychopharmacology and Biological Psychiatry*, *88*, 287–302.
Susser, E., Neugebauer, R., Hoek, H. W., Brown, A. S., Lin, S., Labovitz, D., et al. (1996). Schizophrenia after prenatal famine. Further evidence. *Archives of General Psychiatry*, *53*(1), 25–31.
van der Knaap, N. J. F., Klumpers, F., El Marroun, H., Mous, S., Schubert, D., Jaddoe, V., et al. (2018). Maternal depressive symptoms during pregnancy are associated with amygdala hyperresponsivity in children. *European Child & Adolescent Psychiatry*, *27*(1), 57–64.
Wadhwa, P. D., Culhane, J. F., Rauh, V., Barve, S. S., Hogan, V., Sandman, C. A., et al. (2001). Stress, infection and preterm birth: A biobehavioural perspective. *Paediatric and Perinatal Epidemiology*, *15*(Suppl. 2), 17–29.
Walsh, K., McCormack, C. A., Webster, R., Pinto, A., Lee, S., Feng, T., et al. (2019). Maternal prenatal stress phenotypes associate with fetal neurodevelopment and birth outcomes. *Proceedings of the National Academy of Sciences*, 201905890. https://doi.org/10.1073/pnas.1905890116.
Weinstock, M. (2017). Prenatal stressors in rodents: Effects on behavior. *Neurobiology of stress*, *6*, 3–13.
Wen, D. J., Poh, J. S., Ni, S. N., Chong, Y. S., Chen, H., Kwek, K., et al. (2017). Influences of prenatal and postnatal maternal depression on amygdala volume and microstructure in young children. *Translational Psychiatry*, 7(4), e1103.
Winsper, C., Wolke, D., & Lereya, T. (2015). Prospective associations between prenatal adversities and borderline personality disorder at 11–12 years. *Psychological Medicine*, *45*(5), 1025–1037.
Zagron, G., & Weinstock, M. (2006). Maternal adrenal hormone secretion mediates behavioural alterations induced by prenatal stress in male and female rats. *Behavioural Brain Research*, *175*(2), 323–328.

CHAPTER THREE

The development of stress reactivity and regulation during human development

Melissa L. Engel[a], Megan R. Gunnar[b,*]
[a]Department of Psychology, Emory University, Atlanta, GA, United States
[b]Institute of Child Development, University of Minnesota, Minneapolis, MN, United States
*Corresponding author: e-mail address: gunnar@umn.edu

Contents

Abstract

Adverse experiences during childhood can have long-lasting impacts on physical and mental health. At the heart of most theories of how these effects are transduced into health impacts is the activity of stress-mediating systems, most notably the hypothalamic-pituitary-adrenocortical (HPA) axis. Here we review the anatomy and physiology of the axis, models of stress and development, the development of the axis prenatally through adolescence, the role of experience and sensitive periods in shaping its regulation, the social regulation of the axis at different points in development, and finally conclude with suggestions for future research. We conclude that it is clear that early adversity sculpts the stress system, but we do not understand which dimensions have the most impact and at what points in early development. It is equally clear that secure attachment relationships buffer the developing stress system; however, the

International Review of Neurobiology, Volume 150
ISSN 0074-7742
https://doi.org/10.1016/bs.irn.2019.11.003

mechanisms of social buffering and how these may change with development are not yet clear. Another critical issue that is not understood is when and for whom adversity will result in hypo- vs hyperactivity of stress-mediating systems. These and other issues are important for advancing our understanding of how early adversity "gets under the skin" and shapes human physical and mental health.

Adverse childhood experiences are associated with a wide range of poor physical and mental health outcomes. These experiences become biologically embedded during development. Nearly all of the mechanisms proposed to explain how these experiences "get under the skin" involve the activity of stress-mediating systems (Danese & McEwen, 2012; Del Guidice, Ellis, & Shirtcliff, 2011). McEwen (2000) defines stress as a "real or interpreted threat to the physiological or psychological integrity of an individual that results in physiological and/or behavioral responses" (p. 508). Focusing on psychological aspects, Lazarus and Folkman (1984) define stress as an individual's appraisal of an event as taxing or exceeding his or her resources and thus endangering his or her well-being. A thorough examination of stress consists of three parts—the stimulus event (or input), the perceptual system that processes the stimulus event and appraises the degree of threat posed, and the stress-mediating systems that serve to defend against the event or allow stability through change (allostasis) (McEwen, 2000).

One of the challenges of studying stress developmentally is that all three components undergo changes with development. For example, with regards to the stimulus, McLaughlin and Sheridan (2016) differentiate between two dimensions, threat and deprivation. However, in infancy these two dimensions are indistinct. This is because for the infant deprivation of care is a significant threat to survival and activates many of the stress-mediating systems activated by threat (Leneman & Gunnar, 2018). One reason that development alters how the same adversity is experienced is that the systems that perceive and process threat information change substantially with age, as do the coping skills of the individual. Likewise, the stress-mediating systems undergo marked changes with development and are shaped by their experiences, especially during what appear in studies using animal models to be sensitive periods. A thorough discussion of the development of all three components of stress (input, perception/appraisal, and stress-mediational systems) is beyond the scope of this article. Instead, our goal is to discuss what we currently know about the development of reactivity and regulation of a key stress-mediating system, the hypothalamic-pituitary-adrenocortical (HPA) system, and the role of experiences, particularly those during what might be sensitive periods, in shaping individual differences in how this system responds to stress and influences physical and mental health.

We focus on the HPA axis because it has played a central role in the study of stress and development. It was the first stress-sensitive system shown to be shaped by experiences during a sensitive period in early development in the rat (Levine, 1957). In addition, activity of this system sensitizes and shapes brain systems that are involved in (1) processing and remembering threat, (2) activating the axis and other stress-responsive behavioral and physiological systems, and (3) altering thresholds for terminating stress responding (Herman et al., 2016). We thus begin this chapter by reviewing the anatomy and physiology of the HPA axis and its development. We follow with theories of stress and health with a particular focus on whether the theory is developmental in nature. We continue by considering the role of experience and the possibility of sensitive periods in human development. Then we examine developmental changes in social regulation of stress, a critical feature of stress reactivity and regulation in humans. Finally, we conclude with future directions for research.

1. Anatomy and physiology of the HPA axis and its development

Although we focus on the HPA axis, the sympathetic-adrenomedullary system and, more generally, the autonomic nervous system interact with the HPA axis at many levels to support stress reactivity and regulation. The sympathetic nervous system is relatively well-developed at birth and capable of mounting strong physiological responses to threats to homeostasis (Mulkey & du Plessis, 2019). The parasympathetic arm of the autonomic nervous system is less developed at birth, resulting in poor regulation of autonomic arousal in the first months of life. Development of vagal regulation over infancy and early childhood is believed to play a critical role in emotion regulation and social engagement (Porges & Furman, 2011). In our discussion of the development of the HPA axis, we will note when sympathetic and parasympathetic physiology and development intersect with HPA axis regulation.

The head of the HPA axis lies in the parvocellular neurons in the paraventricular nuclei (PVN) of the hypothalamus that surround the third ventricle of the brain (Herman et al., 2016). These cells produce two releasing hormones, corticotropin-releasing hormone (CRH) and arginine vasopressin (AVP). When produced, these hormones travel through a small portal system from the PVN to the anterior pituitary gland. Here they interact with corticotroph cells, resulting in the release of adrenocortical trophic hormone (ACTH). Very little ACTH is stored in the anterior pituitary, so ACTH has to be cleaved from a large prohormone upon stimulation. ACTH is released into the blood stream where it acts on one of the three zones of the adrenal cortex, the zona fasciculate, to result in the production and release of the steroid hormone cortisol in primates and corticosterone in some other species (Gunnar & Vazquez, 2006).

Once in the blood stream, roughly 90% of cortisol binds to corticosteroid-binding globulin (CBG, transcortin), and when bound is relatively inactive. Plasma assays typically measure both the active and inactive fraction of the hormone in circulation. Because only the unbound fraction can pass through the parotid gland into saliva, salivary cortisol reflects only the active fraction of the hormone, which in many ways makes salivary cortisol a better index of HPA axis activity (Kirschbaum & Hellhammer, 1994). Similarly, cumulative measures of cortisol in hair samples are also believed to reflect the unbound fraction of the hormone (Dettenborn, Tietze, Kirschbaum, & Stalder, 2012). Unbound cortisol is lipid soluble and thus can enter cells without the need for active transport. Once in the cell, cortisol binds to a receptor, creating a hormone-receptor complex that then enters the nucleus of the cell. Once there, cortisol interacts with glucocorticoid receptive elements (GREs) in the promotor regions of many genes, which results in the activation or inhibition of transcription of those genes.

There are two types of receptors that bind cortisol in the brain: mineralocorticoid receptors (MRs) and glucocorticoid receptors (GRs). MRs have a high affinity for cortisol and in the brain are the predominant receptors activated when cortisol is at basal, non-stress levels. GRs become occupied as cortisol rises into stress ranges and/or at the peak of the circadian cycle. (*Note*: there are MRs and GRs in tissues in many parts of the body, but outside the brain MRs bind a salt-water regulating hormone, aldosterone, while GRs bind cortisol). Generally speaking, MRs mediate restorative functions of cortisol and help keep neurons responsive to their neurotransmitters while GRs mediate stress and energy depleting functions (Sapolsky, Romero, & Munck, 2000). There is increasing evidence that the MR (rest, restorative) to GR (stress, depleting) ratio plays a role in mental health (De Kloet, 2014).

Timing is a critical feature in understanding the role of different systems in the stress response. The time between activation of the SAM system to the impact of its catecholamines on organs and tissues is milliseconds: the fight/flight system comes on rapidly and can be quickly terminated. For the HPA axis, the time from stimulation of the PVN to activating or inhibiting gene transcription is slow. It takes roughly 20 min to reach peak cortisol production following activation in the PVN and many more minutes, or even hours, to fully produce the gene-mediated biological effects of the hormone. A rapid non-genomic, cell-membrane mediated impact of cortisol, which appears to operate when cortisol rapidly increases to high levels, has recently been noted (Joëls, Karst, & Sarabdjitsingh, 2018).

The HPA axis is activated by signals transmitted to the PVN by way of the brainstem, in the case of threats to physical homeostasis, or through multisynaptic inputs involving frontolimbic structures, in the case of psychological threat (Herman et al., 2016). Notably, the CRH and AVP cells in the PVN are under constant inhibition; thus, activation requires the inhibition of inhibition and thus inhibitory neurotransmitters. Excitatory neurotransmitter input to the PVN increases inhibition. The HPA axis is under negative feedback regulation whereby increases in cortisol stimulate mechanisms that then inhibit further production. Negative feedback operates at the level of the pituitary and hypothalamus, as well as extra-hypothalamically, in the hippocampus and regions of the prefrontal cortex. Recently, a rapid negative feedback mechanism based on the endocannabinoid system was identified in the hypothalamus, which helps explain inhibition that comes on within minutes of activation of the response (Hill & Tasker, 2012).

In addition to the stress response, basal levels of cortisol follow a diurnal rhythm, with peak levels 30–45 min following morning wakeup and a nadir 30 min after the onset of nighttime sleep (Adam et al., 2017). The diurnal rhythm is critical to healthy functioning and is often disturbed under conditions of chronic stress or in relation to disorders such as depression (Adam et al., 2017). It is also noteworthy that if cortisol is too low for too long, MRs are not fully occupied, and many of the restorative functions they mediate are impaired. On the other hand, if cortisol is too high for too long, the ratio of GR to MR occupation shifts to an overabundance of GR-mediated stress, or energy depleting functions. Thus, the association of cortisol to optimal physical and mental functioning tends to follow an inverted-U in humans and animals alike (Lupien, McEwen, Gunnar, & Heim, 2009). From an evolutionary perspective, enhanced attention and learning in the context of threat heightens likelihood of survival. When chronically or repeatedly activated, however, adverse physical, cognitive, and socioemotional effects are often noted (Lupien, Juster, Raymond, & Marin, 2018). Such effects are often related to patterns of hyper- or hypocortisolism. Whether and when chronic stress leads to hyper- vs hypocortisolism is not well understood.

Which direction the systems shift may relate to the type of stressor (threat vs deprivation), individual differences in genetics, the duration and/or intensity of exposure, and/or the developmental period in which chronic stress was experienced. Critically, blunted cortisol may be specific to the HPA axis, allowing other stress-related systems to exhibit hyperactivity (Koss & Gunnar, 2018). For instance, diminished glucocorticoid modulation of the immune system may enable heightened inflammatory activity under conditions of chronic stress (Miller, Chen, & Parker, 2011).

1.1 Developmental changes in the HPA axis and its regulation

1.1.1 Prenatal

For the first 3 months in utero, the adrenal cortex consists of one large fetal zone that makes dehydroepiandrosterone (DHEA; androgens), which is sent to the placenta and turned into estrogens (Ishimoto & Jaffe, 2011). Two additional zones develop in the middle of the first trimester, which allows cortisol to be produced. While cortisol is briefly evident, the enzyme necessary to produce it is largely inactive until near the end of the second trimester (Ishimoto & Jaffe, 2011). Cortisol generated by the fetus before the late second and third trimester is unlikely to impact brain development, but CRH and cortisol can impact fetal development in other ways earlier on (as reviewed in Leneman and Gunnar (2018)). Notably, whether and how cortisol can impact the developing fetus and child also depends on whether and where there are receptors for the hormone and the type of receptors present. There are complex developmental patterns for MR and GR in the brain that should be taken into account in research on adversity and neurobehavioral development (Rosenfeld, van Eekelen, Levine, & de Kloet, 1993).

As the pregnancy progresses, threats to both the mother and the fetus become more likely to result in elevated cortisol, primarily due to CRH stimulation in the placenta (see review, Gunnar & Davis, 2013). Elevated cortisol in either the mother's plasma or in the placenta is associated with low birthweight, which is in turn linked to hyper-reactivity of the HPA axis, along with vulnerability to cardiometabolic diseases and neurocognitive deficits. Once conceptualized as only a protective organ, the placenta is now thought to also serve as a sensing system that transduces signals to prepare the fetus for the degree of harshness posed by the extra-uterine environment. The notion of fetal programming, or the developmental origins of health and disease (DOHaD), explains how health trajectories, including those mediated by the HPA axis, may be calibrated in utero based on postnatal environmental expectations (O'Donnell & Meaney, 2017). Importantly, although elevations in cortisol may alter fetal developmental trajectories, moderate levels of cortisol are necessary to enable the maturation of critical organs, such as the lungs (Gunnar & Davis, 2013). This may partly explain why levels of unbound cortisol increase in the last trimester.

1.1.2 Infancy and early childhood

Following birth, the adrenal cortex undergoes marked structural changes. The large fetal zone begins to involute and disappears wholly by about 6 months postnatal, allowing the organization of the mature adrenal cortex to emerge. The mature cortex consists of three zones, which differ in the enzymes present and thus the products produced. The glomerulosa produces aldosterone, a salt-water regulating hormone; the fasciculate produces cortisol; and the reticularis produces adrenal androgens, primarily dehydroepiandrosterone (DHEA) (Leneman & Gunnar, 2018). Aldosterone is stimulated by drops in blood pressure and not by ACTH. ACTH, however, stimulates both the production of cortisol and DHEA. DHEA is produced in high volumes in the fetal period, decreases after birth to lower but still significant levels, and then increases again at the beginning of puberty, a phenomenon termed adrenarche. These developmental changes are not, apparently, driven by changes in ACTH and indeed their mechanisms are poorly understood (Quinn, Greaves, Badoer, & Walker, 2018). However, as cortisol and DHEA have opposing effects on many organs and tissues, the higher ratio to cortisol at different points in ontogeny may serve a protective role during development.

Labor and delivery is a major stressor, and newborns are capable of mounting an elevated cortisol response (Leneman & Gunnar, 2018). As noted, in mature individuals, cortisol is typically bound to transcortin. These proteins are low early in infancy but rise throughout the first months of postnatal life. Consequently, total cortisol production is quite low at birth. Yet, because of the low levels of CBG, the unbound and biologically active fraction of the hormone is high at birth and rises even higher when stimulated by stressors such as blood draws, circumcision, and even undressing, weighing, and measuring (Gunnar, Larson, Hertsgaard, Harris, & Brodersen, 1992). Levels also vary with the newborn's state of arousal in the 30 min prior to sampling, with quiet sleep during that time being associated with lowest levels, awake and calm the next lowest, and distressed the highest (Gunnar et al., 1992).

The HPA axis is highly reactive during the first 3 months postnatal, with stressors as mild as a bath producing elevations in cortisol (Gunnar, Talge, & Herrera, 2009). Between 3 and 4 months, fast negative feedback mechanisms develop, resulting in much smaller responses to mild stressors despite the infant crying in response to things such as physical exams (Gunnar & Donzella, 2002). Pain stressors, such as inoculations, continue to produce elevations in cortisol throughout infancy, especially for temperamentally more anxious infants (Gunnar & Donzella, 2002). However, a less marked cortisol response is increasingly observed. As noted, the maturation of fast negative feedback likely allows the response to be more efficient, coupled with the adrenal's decreasing sensitivity to elevations in ACTH (Lashansky et al., 1991). Yet, the most important reason why the axis shows less reactivity to everyday situations that distress infants and young children is that it comes under strong social regulation by caregivers, especially attachment figures (Gunnar & Donzella, 2002).

During infancy and early childhood, the presence of attachment figures is critical to maintaining basal levels of activity and a normal diurnal cortisol rhythm (Gunnar & Donzella, 2002). This is particularly apparent when we examine children who attend full-day out of home childcare. When children start a new childcare, they often show elevations in cortisol that reflect both separation and the novelty of the setting (Ahnert, Gunnar, Lamb, & Barthel, 2004). After a week or so, this response wanes as children adapt to the setting. However, when tracked across the day, by afternoon cortisol levels rise above home levels taken at the same time of day (Vermeer & van IJzendoorn, 2006). The largest increases are observed during the toddler period, with levels across the childcare day approximating those at home by the time the child is approaching 5 years of age (Dettling, Gunnar, & Donzella, 1999). Quality of child care influences these levels (Vermeer & van IJzendoorn, 2006), but it is notable that shortly after children are picked up by their parents at the end of the child care day, they rapidly revert to the low levels typical of late afternoon and evening (Sumner, Bernard, & Dozier, 2010). Thus, separation from the parent during periods of development when the system expects regulation by attachment figures is the most likely explanation for these rising levels at childcare. Whether this daily increase alters neurobehavioral development is unclear. There is evidence, however, that rising levels over the day for more temperamentally fearful children may result in increasing negative emotionality (Gunnar, Kryzer, Van Ryzin, & Phillips, 2011).

Aside from changes in stress reactivity and regulation, the diurnal rhythm develops throughout the first several years of life. Newborns display two peaks in cortisol levels, approximately 12 h apart, which are not coordinated with time of day (Gunnar & Vazquez, 2006). Around 6 weeks, this has matured into one peak in the morning, and, by 3 months of age, one peak in the morning and one valley in the evening. Cortisol patterns correspond to the eating and sleeping patterns of infants and toddlers and do not fully match adult patterns until children stop napping. When this occurs, cortisol tends to fall from morning to afternoon, with a brief post-prandial surge around the noonday meal, as it does in healthy adults.

1.1.3 Childhood and adolescence

Puberty brings on many changes in stress reactivity and regulation. Beginning with adrenarche, at ages 6–8 years, the adrenal increases its production of DHEA. This begins to shift the ratio of DHEA to cortisol. With puberty, cortisol levels do seem to increase, perhaps peaking around age 10–14, or Tanner stage 3 (e.g., Netherton, Goodyer, Tamplin, & Herbert, 2004, but see also Jonetz-Mentzel & Wiedemann, 1993). With age and/or pubertal development, some suggest cortisol responses to laboratory stressors also increase (e.g., Klimes-Dougan, Hastings, Granger, Usher, & Zahn-Waxler, 2001), but this is may be a reflection of increased sensitivity to social evaluation, the key stressor in many of these stress tasks (Sumter, Bokhorst, Miers, Van Pelt, & Westenberg, 2010). Furthermore, sex differences in HPA activity, while less-established during childhood, become more pronounced during adolescence, likely due to bidirectional effects of the hypothalamic-pituitary-gonadal (HPG) and HPA axes (Shirtcliff et al., 2015).

1.2 Brain systems regulating stress and their development

The HPA axis is activated by both ascending information about the state of the body and descending information about anticipation or perception of threat. Ascending information is largely carried by the nucleus tractus solitarius/vagal complex (NTS) and is well-developed at birth (Herman, 2018). With development, higher-order sensory and perceptual processing begins to influence HPA axis reactivity and regulation. As described by Joëls and Baram (2009) the "neuro-symphony of stress" reflects the complexity of the stress system; rather than the HPA axis acting independently, several brain areas and neurotransmitters are highly involved in activating and inhibiting a stress response. Three brain regions with abundant GRs have received the most attention: the amygdala, hippocampus, and medial frontal cortex (Lupien et al., 2009).

1.2.1 *Amygdala*

The amygdala is critical to activation of the HPA stress response (Herman, 2018). However, the amygdala is a complex structure, consisting of three subregions in the rat and even more in the primate (Morgan & Amaral, 2014). In the rat, the central nucleus of the amygdala (CeA) is responsive to systemic stressors, such as inflammation, but not to psychological stressors, such as restraint (Herman, 2018). In contrast, the medial amygdala (MeA) is responsive to psychological stressors but not systemic ones. Both the CeA and MeA impact the PVN through projections to the bed nucleus of the stria terminalis (BNST). Projections from the BNST to the PVN are GABAergic. Thus, as noted earlier, increasing CRH and activating the axis through limbic projections involves disinhibition. In addition to the BNST pathway from the amygdala, the MeA influences the PVN over a preoptic area pathway, while the CeA projects to the NTS. It is also the case that the basolateral amygdala (BLA) is involved in activation of the HPA axis, but its role appears to be to receive and further process threat signals and then pass them on to the CeA or MeA.

The human amygdala increases in size and complexity in the first years of life, with some evidence that the pace of development may be more rapid in girls than in boys (as reviewed in Leneman and Gunnar (2018)). However, whether this relates to the circuits regulating the HPA axis is unknown. There is evidence, again in the rat, that the amygdala is not part of the fear circuit during the early stress hyporesponsive period (SHRP), but traumatic experiences do influence its development (see review, Hostinar, Sullivan, & Gunnar, 2014). As basal corticoid levels rise at the end of the SHRP in the rat, the amygdala becomes part of the fear circuit. During this transition period, however, the dam's presence suppresses the adrenocortical response and blocks amygdala-dependent fear learning. It is unclear whether this is true in human children and, if so, when during development these mechanisms are operating. Recently, Tottenham, Shapiro, Flannery, Caldera, and Sullivan (2019) examined fear conditioning in preschool-aged children and showed that children approached the aversive stimulus when in the parent's presence but not in their absence, similar to the findings in rat pups. As we will discuss in the section on social buffering, the parent's presence in secure relationships powerfully inhibits increases in cortisol to stimuli that, nonetheless, elicit behaviors indicative of distress. Thus, a similar mechanism may operate during the juvenile period in humans. How this relates to activation of the HPA axis, however, is unclear. It may be that by "disrupting" the amygdala-dependent fear circuit, the presence of the parent also "disrupts" connections from the MeA to the BNST and on to the PVN. To our knowledge, this has not been demonstrated.

1.2.2 Hippocampus

The hippocampus exerts a primarily inhibitory influence on the PVN (Herman, 2018). It is involved in regulating the circadian rhythm and responses to psychogenic stressors. Both MR and GR are richly expressed in the hippocampus, with GR being most involved in negative feedback regulation of the axis. The stress response is largely controlled by the ventral hippocampus/ventral subiculum, stimulation of which inhibits activity in the PVN. Outflow from the hippocampal formation travels over the BNST, the medial preoptic area, and the dorsomedial hypothalamus, among other hypothalamic nuclei, before reaching the PVN.

Unfortunately, on a circuit level, we do not know when hippocampal regulation of the PVN comes on-line in human development (Herman, Personal Communication, 2019). It appears to be the case, though, that the subiculum, along with several other hippocampal regions (e.g., CA2) are relatively mature at birth (Jabés, Lavenex, Amaral, & Lavenex, 2011). Other regions experience a protracted period of development. Glucocorticoids play an important role in sculpting the hippocampus. For example, MRs appear to need to be occupied in order for existing and newly generated granule neurons to survive in the hippocampus, while GR occupation, especially with low MR activation, results in neuron loss (Sousa & Almeida, 2002). Thus, both hyper- and hypoactivity of the HPA axis may have significant effects on hippocampal development and the functions it subserves.

1.2.3 Medial prefrontal cortex (mPFC)

The mPFC plays an important role in processing threat and increases in activity in response to numerous psychological stressors (Herman, 2018). This region, which receives extensive projections from other limbic regions, restrains activity of the HPA axis. One subregion of the mPFC is involved in inhibiting the axis in response to psychological stressors (i.e., prelimbic cortex in rodents) while another (i.e., infralimbic in rodents) inhibits the axis in response to systemic stressors and might even play a role in enhancing responses to some psychological stressors (Herman, 2018).

Importantly, unlike the hippocampus, which plays a role in basal levels of ACTH and cortisol, the mPFC seems to only influence the stress response. The connections between the mPFC and the PVN are indirect and involve many of the pathways already described for the hippocampus.

Again, just when in development mPFC regulation of the HPA axis comes on-line is, to our knowledge, unknown. However, if imaging studies of connectivity between the mPFC and amygdala during the processing of threat stimuli are any guide (see Tottenham & Galván, 2016), the restraining role of the mPFC over the HPA stress response in humans may be late-developing, perhaps as late as adolescence. This would allow a long period when childhood adversity could alter the capacity of the developing mPFC to restrain the stress response.

1.2.4 Chronic stress

Most of the work on the impact of chronic stress on the PVN-regulating functions of the amygdala, hippocampus, and mPFC has been conducted in mature organisms. Under conditions of chronic stress, brain structure and function—especially frontolimbic circuitry—are profoundly altered (see for review, Gee & Casey, 2015). Elevated glucocorticoids can result in heightened bottom-up amygdala reactivity and impaired top-down control by the hippocampus and PFC, behaviorally manifested as a shift from reflective to reflexive functioning (Lupien et al., 2009). The inverse actions of these regions, upregulation of the amygdala and downregulation of fronto-hippocampal regions, explain the regional-specific effects of stress on the brain. For instance, changes observed under conditions of chronic stress in rodents include increased CRH expression, dendritic arborization, and reactivity in the amygdala, and atrophy and reduced connectivity in the hippocampus and PFC. Regional specificity is key, as stress-induced neurodevelopmental changes are more resistant to change in some areas than others (Gee & Casey, 2015). The amygdala is both especially sensitive to stress and especially unlikely to recover from its detrimental effects (Tottenham & Sheridan, 2010). On the other hand, recovery is more likely in the PFC and hippocampus, which evince greater plasticity in the absence of stress (as reviewed by Gee & Casey, 2015). Finally, the inverted-U mentioned in the previous section applies to brain development as well in non-human primates, moderate stress has been linked not only to reduced cortisol responses to stress, but to increased prefrontal volume, improved prefrontal function, and reduced anxiety (Parker, Buckmaster, Schatzberg, & Lyons, 2004). Furthermore, these patterns suggest that moderate stress exposure may protect against psychopathology. What is difficult to understand from these animal models is why chronic stress during development in humans is so often associated with hypocortisolism (Bunea, Szentágotai-Tătar, & Miu, 2017), even when the behavioral manifestations are consistent with the rodent and non-human primate models (Gee & Casey, 2015).

2. Theoretical perspectives on stress and development

A number of models have been proposed to describe how early life stress (ELS) impacts physiology and hence mental and physical health. Here we describe several that are most often invoked in the developmental literature. Perhaps the most well-known model, the allostatic load model, is not truly developmental (McEwen & Stellar, 1993). Allostasis describes the function of the stress response, which is to achieve stability through change, or rather, to respond (change) to challenges (stressors) in order to maintain stable functioning. While stress or allostatic adjustments are vital for healthy functioning, they use critical resources and thus incur a cost. If activated frequently or for prolonged periods, the cost or load accumulates, ultimately producing a wear and tear that dysregulates stress-responsive systems and produces mental and physical health problems. While this model is often used to explain the early emergence of diseases of aging with an accumulation of stress over a lifetime, it has also been used to explain why adverse childhood events produce maladaptive childhood trajectories (Danese & McEwen, 2012).

Other models are more explicitly developmental; most of these also posit sensitive periods during which exposure to stressors will have larger impacts. The earliest of these models grew out of research on early handling (Levine, 1957). Brief separations of pups from the dam during a limited period postnatal resulted in a better-calibrated HPA axis and reduced fearfulness. Later it was shown that prolonged, repeated separations and/or low levels of maternal care had the opposite effect (Plotsky & Meaney, 1993). These effects were traced to epigenetic changes in GR genes in brain regions regulating the HPA axis (Zhang, Labonté, Wen, Turecki, & Meaney, 2013). In the rodent, the sensitive period appears to correspond to the first week or so of postnatal life, which would be equivalent to prenatal development in humans. It is unclear whether there is a postnatal sensitive period in human (or more generally primate) development. This will be discussed later in more detail.

Another more elaborated developmental model is the adaptive calibration model (Del Guidice et al., 2011). This model explicitly argues that individual differences in adaption to stressors during development are selected adaptations, even when the behaviors or physiological reactions may seem maladaptive. Developed from a life-history framework, this model argues that life-history relevant dimensions of the environment that cue the predictability/unpredictability of morbidity and mortality influence life-history strategies via activation of the stress response system (neuroendocrine and autonomic systems). The results affect timing of reproduction, number of offspring, rate of growth, risk-taking, and caregiving. Furthermore, the stress response system becomes calibrated and recalibrated to match levels of challenge throughout development. Growing out of another model, the biological sensitivity to context (BSC) model (Boyce & Ellis, 2005), this model argues that a certain degree of threat/challenge is needed in development to program the stress response system to a moderate level of reactivity. Too little challenge and the system remains reactive to even very minor challenges, while too much threat and the system becomes hyper-responsive, over-responding to even mild challenges. In the former case (too little challenge), the highly supportive nature of the caregiving environment fosters highly competent functioning as the individual is able to use this sensitivity in growth promoting ways. In the latter case (too much challenge), the hyper-reactivity is associated with anxiety and high vigilance to threat. If there are extremely high levels of chronic threat during development, however, the stress response system may become hypo-responsive, presumably associated with callous, unemotional personality characteristics. The ACM model describes a number of important sensitive developmental periods in which the stress response system is most likely to shape outcomes.

The idea that some stress during development may be beneficial is not just found in the ACM or in the BSC model that preceded it. Any argument that some stress, or even a good deal of stress, acutely and at particular points in development will have beneficial effects on development is also arguing that the relation between stressful stimulation and developmental outcomes is not linear. If nonlinear, is the relationship U-shaped? If so, then we might expect that a fair amount of stress or adversity would be accommodated in development and would toughen and increase resilience. However, some have suggested that the relationship is more J-shaped, implying that only a modicum of stress or challenge may be needed to organize a well-functioning stress system, while beyond this greater stress begins to be associated with poorer outcomes (Parker & Maestripieri, 2011).

Consistent across most developmental models, but only explicit in some, is the expectation that genetics will set thresholds for how much stress is tolerated before alterations in functioning are observed. The one model that makes this quite explicit is the three hit hypothesis (Daskalakis, Bagot, Parker, Vinkers, & de Kloet, 2013). According to this model, early life stress influences the genetic programming of different systems involved in learning, memory and behavioral adaptation. Vulnerability vs resilience to stress later in life is the result of gene-environment interactions during critical phases in perinatal and juvenile development. The idea of three hits encompasses a genetic vulnerability to disorder and an adverse/stressful early life environment combined with a third hit—a stressful, adverse adult environment. The value to this and several of the earlier models is that stress is not always conceptualized negatively, nor is an adverse early environment expected to always results in poor adult health.

One thing lacking in all of these models are psychosocial protective factors that modify the impact of adversity. The most critical of these are social buffers. While adverse care early in life is a powerful stressor, supportive care and the formation of secure attachment relationships provides a powerful stress buffer that can literally block activation of the HPA axis (Gunnar, 2017; Hennessy, Kaiser, & Sachser, 2009; Hostinar et al., 2014). We discuss the role of social relationships as regulators of stress reactivity and regulation later in this chapter. Here though we include a working model that incorporates social buffering into models of how adversity during development impacts physical and mental health (see Fig. 1, adapted from Koss & Gunnar, 2018). According to this working model, early adversity consists of conditions that threaten organism viability, either because of the absence of nurturing stimulation or the presence of harmful stimulation. The effects of these adverse conditions are modified by the individual's genetic endowment and the presence and availability of supportive caregivers, most powerfully the people with whom the child has formed an attachment relationship. The activation of the HPA axis as it interacts with other stress-mediating systems impacts the development of physical and mental health over the course of development, but perhaps most importantly during sensitive periods when the stress system and the systems underlying the physical and mental outcomes are developing and thus highly plastic. This model builds on many of the ideas in earlier models, including the "third hit" model (Daskalakis et al., 2013), the importance of sensitive periods for calibration and recalibration (Del Guidice et al., 2011), and the argument that in the presence of supportive caregivers (Gunnar, 2017), exposure to a moderate amount of stress and challenge supports healthy development (Parker & Maestripieri, 2011).

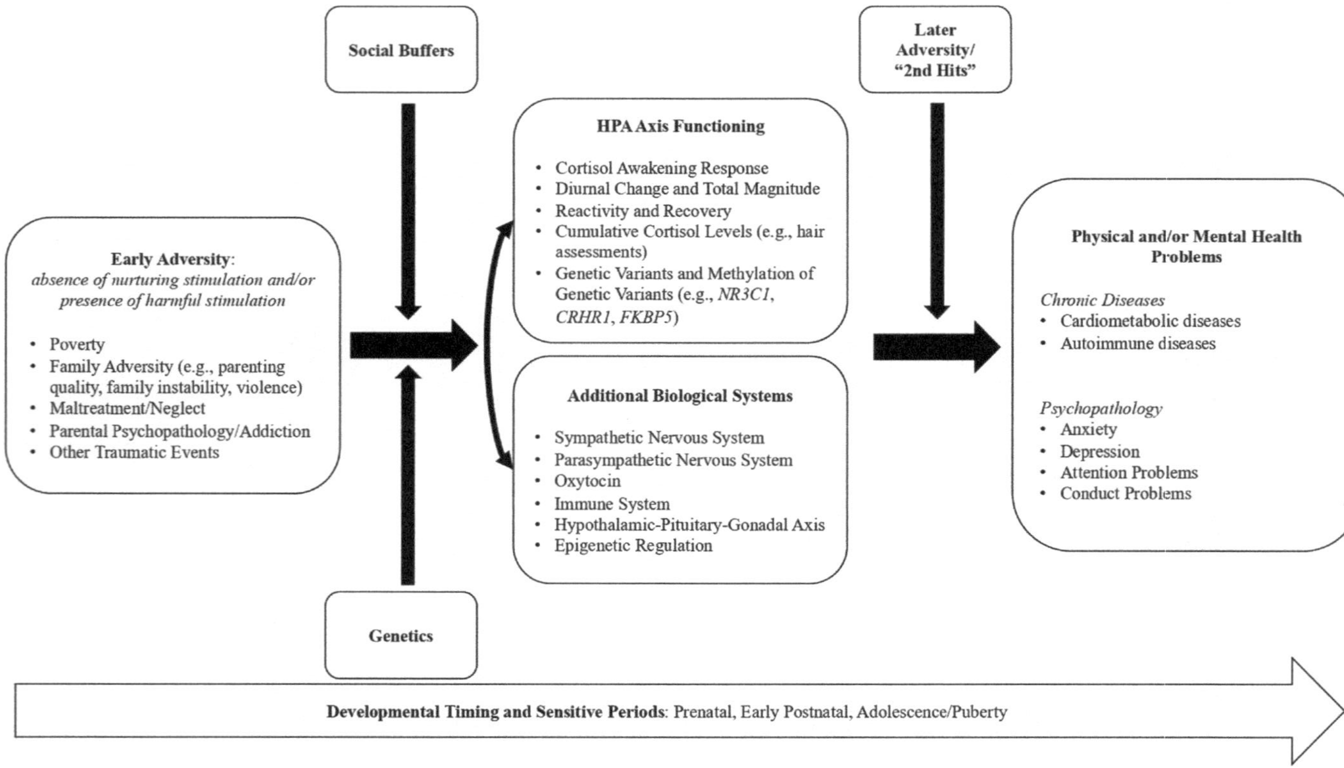

Fig. 1 Working model of the relationship between early adversity, the HPA axis, and physical and mental health problems. According to this working model, early adversity consists of conditions that threaten organism viability, either because of the absence of nurturing stimulation or the presence of harmful stimulation. The effects of these adverse conditions are modified by the individual's genetic endowment and the presence and availability of supportive caregivers, most powerfully the people with whom the child has formed an attachment relationship. The activation of the HPA axis as it interacts with other stress-mediating systems impacts the development of physical and mental health over the course of development, but perhaps most importantly during sensitive periods when the stress system and the systems underlying the physical and mental outcomes are developing and thus highly plastic. *Figure adapted from Koss, K. J., & Gunnar, M. R. (2018). Annual research review: Early adversity, the hypothalamic-pituitary-adrenocortical axis, and child psychopathology.* Journal of Child Psychology and Psychiatry, *4(59), 327–346.*

3. Sensitive periods

Most of our knowledge of sensitive periods for shaping the HPA axis comes from work in the rat. In rat models, repeated prolonged separations, low maternal care, and abusive care have all been associated with shaping hyper-responsive ACTH and corticosteroid responses to stressors. The effect of these early life manipulations on the axis appear to be due to changes in MR:GR ratios in the hippocampus and mPFC (Oitzl, Champagne, van der Veen, & de Kloet, 2010). In rats, the sensitive period appears to be in the first weeks of life, although there may be another sensitive period in the peripubertal period (Romeo, 2010). Importantly, sensitive periods in animal models can be determined with some accuracy, as the timing and type of adversity is experimentally controlled. Unfortunately, assessing sensitive periods in humans is more complicated because we have to rely on correlational designs or experiments of nature. In addition, studies in humans often use retrospective reports of adversity, which are of dubious accuracy (Reuben et al., 2016). For these reason, we need to be cautious in interpreting data on sensitive periods in humans.

3.1 The prenatal period

The Barker hypothesis or more generally the Developmental Origins of Health and Disease (DoHAD) model asserts that health trajectories are roughly established in utero as the fetus prepares for whatever environment it expects to encounter (O'Donnell & Meaney, 2017). According to the Barker hypothesis (Barker & Osmond, 1986), intrauterine stress reduces the size of the body, including the nephrons, elevating risk for high blood pressure, and also increases reactivity of the HPA axis. A number of studies have found that low birthweight, a proxy measure of prenatal stress, is associated with higher fasting plasma cortisol and increased reactivity of the HPA axis to psychosocial stressors (e.g., Carpenter, Grecian, & Reynolds, 2017), although this effect may be specific to females. This may be because in response to maternal stress, the female more than the male placenta increases its permeability to maternal cortisol by larger reductions in the expression of 11β-hydroxysteroid dehydrogenase (11β-HSD) enzymes (Clifton, 2010). While this sex difference might increase female vulnerability to later stress-associated affective pathologies (i.e., depression), it may also have protective effects. According to the sex-based viability-vulnerability hypothesis (Sandman, Glynn, & Davis, 2013), female fetuses adapt more than male fetuses to harsh intrauterine environments, notably slowing their growth, which increases their chance of surviving difficult gestations in which nutritional resources are scarce.

Prenatal stress effects on HPA axis regulation likely involve impacts of prenatal glucocorticoids on the hippocampus. This has been demonstrated in non-human primates using prenatal injections of dexamethasone which, even at the lowest doses, result in decreased neuronal numbers in many hippocampal regions (Uno et al., 1990). Other mechanisms, including maternal inflammation, likely play a role as well (Nazzari et al., 2019). Maternal nutrition is an important factor, with undernutrition leading to elevated HPA activity and decreased GR expression. This in turn results in decreased 11β-HSD2 levels and consequently increased fetal exposure to glucocorticoids. Fetal lipogenesis increases, along with expressions of 11β-HSD1 and GR in maturing adipose tissue, ultimately programming the later development of metabolic syndrome (Correia-Branco, Keating, & Martel, 2015). Maternal use of drugs and alcohol may also alter HPA axis functioning. However, findings are presently inconsistent, with some reporting hypocortisolism and others reporting hypercortisolism (as reviewed by Fisher, Kim, Bruce, & Pears, 2012).

Although there is growing evidence that prenatal stress shapes HPA axis reactivity and regulation, less clear is whether it results in hyper- or hypoactivity. A recent meta-analysis found equivalent effect sizes for programming patterns of both hypo- and hypercortisolism (Pearson, Tarabulsy, & Bussières, 2015). What accounts for these two patterns is unclear, but it may in part be related to the timing of the stressors. Several mechanisms, which change throughout gestation, regulate the impact of maternal HPA axis activity on the developing fetus. For instance, early in gestation, 11β-HSD2 enzyme renders cortisol inert, in order to protect the fetus from high levels of maternal cortisol. Later in gestation, however, the fetus begins producing its own cortisol, maternal cortisol rises, and 11β-HSD2 expression decreases, potentially allowing maternal stress to pass to the fetus and engender effects on physical growth and brain maturation (Ishimoto & Jaffe, 2011).

3.2 The early postnatal period/infancy

Human studies of ELS often focus on maltreatment. However, it is difficult to use maltreatment to isolate the timing of early adversity, as maltreatment is often not limited to one age period. In studies of non-human primates, though, the timing can be more explicitly determined. Furthermore, cross-fostering designs in these studies allow the effects of experience to be differentiated from genetics. This work shows that rhesus monkeys maltreated in the first 3 months of life display heightened cortisol, which in turn predicts decreased white matter integrity in the corpus callosum, among other brain regions, in adolescence (Howell et al., 2013). Furthermore, as juveniles, both left and right amygdala volumes have been positively associated with the degree of abuse in infancy, potentially predisposing heightened stress reactivity (Howell et al., 2014).

In humans, evidence generally suggests that early maltreatment is associated with decreased hippocampal volume and increased amygdalar volume—potentially indicative of increased stress reactivity and impaired regulation—later in life, but a sensitive period has yet to be well-delineated (e.g., Hanson et al., 2015). In the one published study examining stress timing and HPA activity, traumatic exposure prior to age one was associated with slower recovery from a stress task, suggestive of deficits in negative feedback (Kuhlman, Vargas, Geiss, & Dopez-Duran, 2015). Importantly, psychopathology is highly prevalent in individuals with a history of maltreatment and may moderate the relationship between maltreatment and HPA activity (Koss & Gunnar, 2018). For instance, one study found that children who experienced physical and sexual abuse before age 5 but not afterward displayed altered cortisol levels over the day, but only if they also exhibited elevated internalizing problems (Cicchetti, Rogosch, Gunnar, & Toth, 2010). This raises the possibility, consistent with the three hit model, that both a genetic vulnerability to anxiety and depression, plus early life stress are needed to alter HPA axis development postnatally. In sum, while the maltreatment literature suggests that the HPA axis is implicated, it has yet to provide substantial evidence for sensitive period effects.

The timing of maternal depression (active vs in remission) has also been studied in relation to shaping of the developing axis. There is some evidence that maternal depression in the infant's first year has lasting effects on the axis (e.g., Halligan, Herbert, Goodyer, & Murray, 2004), yet other research suggests these effects come with maternal depression later in development or are cumulative (see review, Koss & Gunnar, 2018). One challenge in using the literature on maternal depression to try to identify sensitive periods is that families with more resources may be able to partially buffer the child during maternal depressive episodes. Additionally, even during periods of remission, the mother-child relationship may continue to be stressful.

The best evidence for early postnatal sensitive periods comes from studies of children who experience early institutional rearing followed by adoption into supportive homes. Rearing in institutions (e.g., orphanages) is a significant stressor which typically results in poor physical, social, and cognitive development (van Ijzendoorn et al., 2011). When children are placed in families, they experience marked improvement in both physical and emotional care and respond by exhibiting massive developmental catch-up (van Ijzendoorn & Juffer, 2006). One study of Romanian adoptees reported no effect on wake up or evening cortisol levels for children adopted at birth or in the first months after birth, but elevated levels for those adopted after 8 months, with differences persisting for years following adoption (Gunnar, Morison, Chisholm, & Schuder, 2001). Another study of Romanian children showed that even 20 years after adoption, those individuals adopted under 6 months were more similar to individuals who had never been in institutional care, while those adopted after 6 months of age showed continuing dysregulation of the axis (Kumsta et al., 2017). Both of these studies involved individuals who experienced some of the harshest institutional care, as these adoptions occurred soon after the fall of the communist regime in Romania. Thus, the fact that individuals adopted under 6 months did not differ from never-institutionalized individuals is remarkable. Other studies of the diurnal rhythm in children adopted from institutions have not included individuals adopted as early as in these two studies. We have examined the increase in cortisol after awakening in 7- to 15-year-old children adopted from institutions in a variety of countries (Leneman, Donzella, Desjardins, Miller, & Gunnar, 2018). Those adopted over 16 months of age (our median) showed smaller increases following awakening, while those adopted earlier showed a response similar to the comparison youth. To our knowledge, other studies of the diurnal rhythm in previously-institutionalized youth have not addressed the question of sensitive periods.

Of greater relevance to the sensitive period question, however, are studies examining reactivity of the axis to stressors. This is because sensitive period effects in animal models are typically reported for reactivity and not home cage levels of ACTH and glucocorticoids (Sánchez, Ladd, & Plotsky, 2001). Unfortunately, to our knowledge there has been only one study of reactivity that has addressed the sensitive period question. Nonetheless, this was an important one, as it involved random assignment and thus permits us to conclude that there was a causal role of institutional care on the HPA axis. In the Bucharest Early Intervention Study, children from several institutions in Bucharest, Romania were randomly assigned to remain in care as usual or be placed in foster care designed and supported by the study. When the children were an average of 12 years, they were administered the Trier Social Stress Test (TSST). At the time of random assignment, some of the children were under 2 years of age (the median age) while others were older. The results showed that those who were over 2 years when placed in foster care and those who remained in care as usual had severely blunted cortisol and autonomic responses to the stressor task, but those under 2 years when placed in foster care and those never in institutional care showed normal responses to the task. A similar result was obtained in a study of social buffering. In that study, for children adopted under 18 months (the median), the mother's presence buffered the cortisol response to the TSST; for those over 18 months, the mother's presence or absence had no effect (Hostinar, Johnson, & Gunnar, 2015). Thus, there is some suggestion that, at least with regards to institutional deprivation, the postnatal sensitive period may close sometime between about 1.5 and 2 years. More studies are needed, however, to be sure this is the case.

3.3 Pubertal recalibration/adolescence

Puberty may open another window of plasticity for the HPA axis. Indeed, puberty is a time of heightened plasticity in neurobiology, which makes it a period of both risk and opportunity (Blakemore, Burnett, & Dahl, 2010). Animal studies suggest it is the peripubertal period per se, as opposed to age, that enables heightened neural plasticity during this time, and growing evidence supports this hypothesis in humans (Blakemore et al., 2010). Adolescence is also a period of heightened amygdala response to emotional stimuli (Hare et al., 2008). Moreover, as discussed earlier, basal levels of cortisol increase with puberty (Gunnar & Vazquez, 2006). The greater sensitivity to social evaluative threat in adolescence may heighten anticipatory cortisol and autonomic reactions to stressors like public speaking tasks. As evidence, peripubertal adolescents display less variability in anticipatory cortisol than do postpubertal adolescents (Sumter et al., 2010).

Along with heightened activity of the HPA axis, puberty may increase the brain's sensitivity to stress. For example, the amygdala, hippocampus, and striatum, which are all sensitive to stress, increase in volume across adolescence (Gee & Casey, 2015). Also, GR and mRNA in the PFC are significantly elevated in adolescence, compared to infancy and young adulthood (Lupien et al., 2009). It is likely that stress effects will become sexually dimorphic with puberty. At some point in development, HPA axis reactivity to stressors like the TSST begin to follow the menstrual cycle in females, with females showing smaller responses than males in some phases and responses similar to males in others (Kirschbaum, Kudielka, Gaab, Schommer, & Hellhammer, 1999). It is thought that the increase in affective pathology in adolescence, as well as the emergent sexual dimorphism in stress sensitivity may be partially due to HPA-HPG interactions (Sisk & Zehr, 2005).

If puberty opens a period of heightened plasticity of the brain in response to stressors, does it also have this effect on the HPA axis? Studies in rats suggests this is the case. In adulthood, rats who had been exposed to stressors during the peripubertal period showed elevated, poorly regulated corticosterone patterns, compared to those who had been exposed as young adults (Romeo, 2010). Relative to adulthood, stressors in adolescent rats were associated with protracted activation of the HPA axis; with an immature negative feedback system, peripubertal rats take longer to mount elevations in glucocorticoids and longer to return to baseline (Lupien et al., 2009). Likely contributing to such long-lasting effects in rats, stressful experiences during adolescence increase dendritic complexity in the amygdala while decreasing dendritic complexity in hippocampus and PFC (Romeo, 2017).

If there is a second sensitive period during puberty for shaping the HPA axis and brain regions that help regulate it, then this likely cuts both ways. Stressors during this period may have greater long-term effects on the axis and associated brain regions, but improved conditions during this time may allow remodeling of systems shaped during early development by harsh conditions. This is known as the pubertal stress recalibration hypothesis (Koss & Gunnar, 2018). Three recent human studies suggest than the negative physiological effects of early life stress may be reversed throughout puberty (DePasquale, Donzella, & Gunnar, 2019; Gunnar & Quevedo, 2007; King et al., 2017). One of these studies examined internationally-adopted youth who experienced early life adversity in the form of institutional rearing, followed by adoption into supportive families, compared to non-adopted controls. While early-pubertal previously-institutionalized youth displayed patterns of hypocortisolism, children at later stages in puberty did not. This suggests that the maladaptive pattern of blunted cortisol may normalize throughout puberty (DePasquale et al., 2019).

4. Social regulation of stress during development

Humans evolved to be social animals; consequently, the presence of others plays an important role in regulating stress. Those with whom we have close relationships can be major sources of stress, as well as our most powerful stress buffering resources (Beckes & Coan, 2011; Hennessy et al., 2009; Hostinar et al., 2014). Social buffering describes the presence (real or representational) of a social partner decreasing the activity of a stress-mediating system, such as the HPA axis, in the context of threat (Gunnar, 2017). As with any developmental phenomenon, social buffering changes form overtime. We hypothesize that the relationship with an attachment figure early in life not only serves to buffer stress in infancy and early childhood, but also to provides a framework for social buffering by peers and romantic partners across the life course (see Gunnar, 2017 for review). This also means that disruptions in early attachment relationships may impair mechanisms through which individuals are able to use others to regulate their physiological stress response system.

4.1 Infancy

One reason that adverse care early in life has powerful effects on development is that a child's primary caregivers are an external extension of their internal stress regulatory system. Under many conditions, the parent's presence permits the child to emit distress signals, which are critical to alerting the parent to the child's needs, without simultaneously increasing the production of stress-mediating hormones (see Gunnar, 2017 for review). Much of the theorizing about social buffering by primary caregivers is grounded in attachment theory. Bowlby (1969) argued that the attachment figure is embedded as part of the infant's stress and emotion regulatory system. When the child is deprived of an attachment figure (as in institutional care), or when the attachment figure(s) is unable to provide sensitive and responsive care, the child is less able regulate stress.

The stress buffering role of the caregiver is exhibited across mammalian species (Hennessy et al., 2009). In rat pups, stimuli by the dam maintains the stress hyporesponsive period lasting approximately 7–9 days postnatal (see review, Hostinar et al., 2014). Following this period, the rat HPA axis is responsive to threat, except in the presence of the dam (Santiago, Aoki, & Sullivan, 2017). Exactly when in human development contact with caregivers becomes capable of reducing the HPA response to stressors is unclear.

In premature infants, a kangaroo care intervention was shown to alter numerous aspects of newborn regulation and, when these children were followed up, greater contact with the mother in the hospital nursery was predictive of better regulation of cortisol and oxytocin (OT) (as reviewed in Feldman (2015)). Thus, it seems likely that skin-to-skin contact is an effective stress regulator from birth. Nonetheless, during more typical parent-infant interactions, the HPA axis remains highly reactive even to mild stressors, like being undressed and handled, during the first 3 months. However, sensitive care serves to bring the system back to baseline more rapidly during this time (Albers, Riksen-Walraven, Sweep, & de Weerth, 2008). Likewise, sensitive care from a new, unfamiliar caregiver can reduce the cortisol response to separation in infants as late as 9 months of age (Gunnar et al., 1992).

At least as early as 1 year of age, the presence and availability of the attachment figure, even in the absence of physical touch, is able to buffer the HPA axis (see for review, Gunnar, 2017). By this age, too, the quality of the relationship makes a difference, with the presence of the attachment figure in a secure relationship providing a more powerful buffer than an attachment figure in an insecure relationship (Gunnar, 2017). In the second year of life, this has been noted in studies using childhood inoculations, scary toys, and being familiarized with a child care center prior to starting child care. While having the presence of the attachment figure buffer stress in secure relationships is likely important for all young children, it appears to be especially important for temperamentally fearful children (Gunnar, 2017). In addition to dampened HPA activity, sensitive and responsive caregiving is reflected in increased left frontal electrical activity (related to approach and positive emotionality) and increased vagal tone (Hane & Fox, 2006). Together, these socially-induced biological effects protect the developing child from negative effects of everyday stressors while allowing them the opportunity to explore and gain coping skills.

Secure attachment relationships early in life may also buffer children from more significant stressors. For example, a recent report demonstrated that cortisol levels during a well-child pediatric visit were a function of both the family's income and the parent-child attachment relationship (Johnson, Mliner, Depasquale, Troy, & Gunnar, 2018). For toddlers living in poverty, cortisol levels were high across the visit if the attachment relationship was insecure, but not if it was secure. Similarly, a recent longitudinal study was conducted in an impoverished area of Cape Town, South Africa, in which attachment security was assessed in infancy and the TSST was performed at 13 years of age (Fearon et al., 2017). In this study, growing up in poverty resulted in a blunted HPA response to the TSST if the children had been insecurely attached in infancy, but not if they were securely attached. Thus, attachment security protected the axis from being adversely programmed by poverty.

4.2 Childhood

Secure attachment relationships continue to play a critical role in buffering children from adversity throughout childhood. For example, children growing up in a war zone whose mothers were observed to provide sensitive containment of the child's fear were less likely to develop PTSD and exhibit dysregulation of the HPA and sympathetic system (as described in Feldman (2015)). Evidence also suggests that caregivers continue to effectively buffer stress throughout middle childhood. Researchers employed the TSST for Children (TSST-C) with girls aged 7–12 years using three post-stress recovery conditions: having full access to mom, talking to mom on the phone, or being without any contact with mom (Seltzer, Ziegler, & Pollak, 2010). Girls returned to baseline cortisol fastest when mom was physically present, followed by talking to her on the phone and then the alone condition. Moreover, urinary oxytocin levels were higher in both maternal conditions than when alone. Oxytocin is a peptide thought to play a pivotal role in the social regulation of the stress system by both decreasing HPA activity and supporting the parasympathetic nervous system (Feldman, 2017). Recent work on the neurobiological basis of attachment, coupled with experiments with children adopted from orphanages, suggests there may be a sensitive period for the development of oxytocin-dopamine connections which exert enduring effects on the neurobiology of social relationships, including their ability to physiologically buffer stress (see for review, Feldman, 2017).

Peer relations become increasingly important to children during the early and middle childhood years. "Are you my friend?" is a dominant topic of conversation among preschoolers. As early as 3–5 years of age, peer rejection is associated with elevated levels of cortisol (Gunnar, Sebanc, Tout, Donzella, & Van Dulmen, 2003). Young children tend to show an increase, rather than the normal decrease, in cortisol across full child care day; however, it has recently been shown that a randomized social skills training intervention resulted in a return to the normal diurnal decrease among preschoolers (Larose et al., 2019). This suggests that part of what drives the HPA axis in full-day child care is the stress of trying to maintain relations with other children when none of the children possess the social skills to do so predictably and controllably. Children who are deficient in social skills are often the targets of bullying, and there is good evidence that child characteristics, including genotype, play a role in whether a child is chronically bullied (Bowes et al., 2013). Thus, genetically informed designs may be needed to sort out the effect of bullying on child functioning, including functioning of the HPA axis. In a study of bullying using a monozygotic twin design, the more victimized twin showed a diurnal cortisol pattern reminiscent of PTSD (Brendgen et al., 2017). In contrast, when responses to the TSST were examined in middle childhood, bullied children exhibited larger cortisol responses than non-bullied children (Chen, Kong, Deater-Deckard, & Zhang, 2018).

If other children can be a source of stress, can they be a source of stress buffering and, if so, under what circumstances? There is remarkably little research on these questions. One very relevant study, however, suggests that at least by middle childhood best friends can help buffer against bullying by other children (Adams, Santo, & Bukowski, 2011). Children reported on positive and negative experiences with peers over several school days and also collected saliva for cortisol analyses. Cortisol levels were lower 20 min after a negative encounter if the children reported that their best friend was with them at the time.

4.3 Adolescence

While adolescents continue to both report and benefit from social support and emotional closeness with their parents, the power of attachment figures to physiologically buffer stress decreases in adolescence (Gunnar, 2017). In one study, parents were able to effectively buffer cortisol responses to a social evaluative stressor among 9 and 10 year old children, but not among 15 and 16 year old children (as reviewed in Gunnar (2017)). In this paradigm, parents were present while the child or adolescent prepared their speech in the TSST, but were not there during speech delivery or the math portion of the task. We do not know whether there would be a similar loss of stress buffering effectiveness if the parent was present during recovery after the task, as in previously described study of Seltzer et al. (2010). Nor do we know if the parent's presence fails to buffer the adolescent when the stressor is a physically threatening event, such as a medical procedure.

What we know, however, is that for a social evaluative stressor, having the parent present during speech preparation becomes less effective with pubertal stage, rather than age, during the transition to adolescence (as reviewed in Gunnar (2017)). It is possible that this shift in the effectiveness of parental social buffering is part of a larger shift in stress reactivity and regulation during the pubertal period that was discussed earlier. If as discussed, the HPA axis become more reactive during puberty, and changes with puberty make the parent or attachment figure a less powerful stress buffering agent, this might partially explain the increase in affective psychopathology that arises during early adolescence (Koss & Gunnar, 2018). It is notable that the reduction in parental social buffering potency in adolescence may be seen on the neural level as well. For example, in one study, children and adolescents completed an fMRI task, in which they viewed a photo of the face of either their mother or a stranger. The mother's face successfully buffered elevated amygdala activity in children, but not in adolescents (Gee et al., 2014). Given our pervious discussion of the amygdala's role in activating the HPA axis, these results further support adolescence as a time of heightened stress vulnerability. There are likely individual differences in the ability of parents and friends to serve as stress buffers, which may be dependent upon early and current relationship quality, sex, and other factors. For instance, in one study, relationship quality between 16-year-olds and their mothers predicted their ability to use holding the hand of a friend to decrease neural activity in response to electric shocks at age 24, demonstrating the lasting physiological effects of early attachment relationships (Coan, Beckes, & Allen, 2013).

If the parent becomes less potent of a stress buffer, does the role of stress buffer shift to best friends? It is not clear. This does not seem to be the case when adolescents are preparing their speech in the TSST. Here, the presence of a best friend to provide social support actually results in larger cortisol responses to the TSST, even though the friend is not there while the youth are actually performing and being judged (as reviewed in Gunnar (2017)). Perhaps their presence heightens concerns about social evaluation. It is unclear, however, whether the presence of a friend might buffer stress if the friend was also experiencing social evaluation. This might be more like the idea of "sharing the load" in the Beckes and Coan (2011) Social Baseline Theory. The quality of the friendship likely matters to whether friends can regulate stress. In one study, after a speech stressor friends followed a script to ask the target youth a set of questions about the task (Calhoun et al., 2014). Cortisol levels were higher and remained higher as a function of low vs high friendship quality.

In sum, attachment figures serve as extremely powerful external regulators of children's physiological stress, beginning in infancy and extending throughout childhood. Peers and later friends play important roles in children's lives beginning in the preschool years. While it is clear that being bullied by other children is a source of stress that affects children's emotional well-being and the activity of stress-mediating systems, we know much less about whether, how, and when peers and friends can serve as social stress buffers. As will be discussed, this is an important future direction for research.

5. Conclusions and future directions

There is now convincing evidence that early adversity sculpts the stress system, although there is as great a deal that we still do not understand. First, we do not understand the dimensions of early adversity and how they intersect with the development of the nervous system and thus with stress reactivity and regulation. Second, it is clear that secure relationships buffer the developing stress system, supporting its regulation. However, we do not know if these relationships buffer the system from all stressors or only from psychosocial stressors. Third, while it may be impossible to prove the existence of sensitive periods that set the tone of the stress system in humans, we still need a better understanding of when during development the stress system is more or less plastic and open to the impact of experience. As reviewed, the prenatal period, postnatal period perhaps to as late as 2 years, and again the peripubertal period may be when stress-mediating systems are calibrated and recalibrated during human development. While evidence is accumulating, the entire topic of sensitive periods needs further study.

Fourth, is very clear now that when we see an altered pattern of HPA axis reactivity, it is as likely to be hypo-reactivity as hyper-reactivity. It is very unsatisfying to argue that exposure to adversity will result in a change in functioning, without being able to predict the direction of change. Yet, we still have very little understanding of whether and when to predict increased production vs decreased production of cortisol, nor do we have a good handle on how high vs low cortisol levels affect the developing nervous system.

Fifth, although in order to drill down we focused this chapter primarily on the HPA axis, this axis is, of course, not the whole story. In order to fully understand the development of stress and the impact of stressors on development we need a broader lens that includes other stress-mediating systems (Koss & Gunnar, 2018). Interaction of the HPA, autonomic and immune systems are perhaps the most critical to incorporate into an integrated model. As we expand our work to include multiple systems, timing on multiple scales becomes important. Likewise, we need to understand the cascading and compensatory processes of stress experienced on different time scales from minutes to hours to days, and into different periods of development.

Sixth, we have spent almost no time in this chapter discussing sex differences. In part this is because there is little consistent evidence of sex differences in how stress-mediating systems respond to stressors prior to adolescence. Once in adolescence, there is still relatively little work on sex differences in stress. This must change, however, and is a very fruitful area for future research.

Seventh, as in all other areas of development, we need to understand heterogeneity. There are marked individual differences in stress reactivity and regulation and in the impact of stress physiology on the developing brain. Some children are much more reactive and show larger impacts of exposure to adversity; others are more resilient. Genetics as well as experience likely influence this heterogeneity. With regards to genetics, we need to examine whether we are dealing with a stress-diathesis model, such that particular genotypes result in greater sensitivity to the environment and hence a more altered stress axis. Alternatively, particular genotypes may result in greater sensitivity, which can mean a more resilient axis under supportive conditions and more dysregulated axis under adverse conditions (see Differential Susceptibility Hypothesis, Belsky, 2016).

Finally, for better or for worse, social relationships are critical to stress reactivity and regulation. We have learned a good deal about social buffering and social sources of stress in human development. We still do not understand the neurobiology that regulates social buffering and its changes with development. Because the flip side of buffering in very young children is the stress of inadequate social care, a better understanding of how the nervous system processes social relationships and how it uses them to regulate the activity of stress-mediating systems will have a twofold benefit: we will understand more about how neglectful relationships can harm development and how healthy relationships foster resilience.

In all, 61 years after Seymour Levine published his first paper on the impact of early handling on the HPA axis in rat pups (Levine, 1957), research on the role of experience in shaping the axis in humans is still growing. Our theories of how early life stress impacts development still place the shaping of stress-responsive systems as the core mechanisms. There is still a great deal to learn, with great expectations of what research in the near future will yield.

Acknowledgments

This chapter was written with support to the second author from the JPB Foundation through a grant to the JPB Research Network on Toxic Stress: A Project of the Center on the Developing Child at Harvard University, the Canadian Institute for Advanced Research, Child and Brain Development program, and HD075349.

References

Adam, E. K., Quinn, M. E., Tavernier, R., McQuillan, M. T., Dahlke, K. A., & Gilbert, K. E. (2017). Diurnal cortisol slopes and mental and physical health outcomes: A systematic review and meta-analysis. *Psychoneuroendocrinology*, *83*, 25–41.

Adams, R. E., Santo, J. B., & Bukowski, W. M. (2011). The presence of a best friend buffers the effects of negative experiences. *Developmental Psychology*, *47*(6), 1786–1791.

Ahnert, L., Gunnar, M. R., Lamb, M. E., & Barthel, M. (2004). Transition to child care: Associations with infant-mother attachment, infant negative emotion, and cortisol elevations. *Child Development*, *75*(3), 639–650.

Albers, E. M., Riksen-Walraven, J. M., Sweep, F. C. G. J., & de Weerth, C. (2008). Maternal behavior predicts infant cortisol recovery from a mild everyday stressor. *Journal of Child Psychology and Psychiatry, and Allied Disciplines*, *49*(1), 97–103.

Barker, D. J. P., & Osmond, C. (1986). Infant mortality, childhood nutrition, and ischaemic heart disease in England and Wales. *The Lancet*, *327*(8489), 1077–1081.

Beckes, L., & Coan, J. A. (2011). Social baseline theory: The role of social proximity in emotion and economy of action. *Social and Personality Psychology Compass*, *5*(12), 976–988.

Belsky, J. (2016). The differential susceptibility hypothesis: Sensitivity to the environment for better and for worse. *JAMA Pediatrics*, *170*(4), 321–322.

Blakemore, S. J., Burnett, S., & Dahl, R. E. (2010). The role of puberty in the developing adolescent brain. *Human Brain Mapping*, *31*(6), 926–933.

Bowes, L., Maughan, B., Ball, H., Shakoor, S., Ouellet-Morin, I., Caspi, A., et al. (2013). Chronic bullying victimization across school transitions: The role of genetic and environmental influences. *Development and Psychopathology*, *25*(2), 1–26.

Bowlby, J. (1969). Attachment and loss: Attachment. In Vol. 1. New York: Basic Books.

Boyce, W. T., & Ellis, B. J. (2005). Biological sensitivity to context: I. An evolutionary–developmental theory of the origins and functions of stress reactivity. *Development and Psychopathology*, *17*, 271–301.

Brendgen, M., Ouellet-Morin, I., Girard, A., Lupien, S., Vitaro, F., Dionne, G., et al. (2017). Environmental influence of problematic social relationships on adolescents' daily cortisol secretion: A monozygotic twin difference study. *Psychological Medicine*, *47*(3), 460–470.

Bunea, I. M., Szentágotai-Tătar, A., & Miu, A. C. (2017). Early-life adversity and cortisol response to social stress: A meta-analysis. *Translational Psychiatry*, 7(1274), 1–8.

Calhoun, C. D., Helms, S. W., Heilbron, N., Rudolph, K. D., Hastings, P. D., & Prinstein, M. J. (2014). Relational victimization, friendship, and adolescents' hypothalamic–pituitary–adrenal axis responses to an in vivo social stressor. *Development and Psychopathology*, *26*(3), 605–618.

Carpenter, T., Grecian, S. M., & Reynolds, R. M. (2017). Sex differences in early-life programming of the hypothalamic-pituitary-adrenal axis in humans suggest increased vulnerability in females: A systematic review. *Journal of Developmental Origins of Health and Disease*, *8*(2), 244–255.

Chen, G., Kong, Y., Deater-Deckard, K., & Zhang, W. (2018). Bullying victimization heightens cortisol response to psychosocial stress in Chinese children. *Journal of Abnormal Child Psychology*, *46*(5), 1051–1059.

Cicchetti, D., Rogosch, F. A., Gunnar, M. R., & Toth, S. L. (2010). The differential impacts of early physical and sexual abuse and internalizing problems on daytime cortisol rhythm in school-aged children. *Child Development, 81*(1), 252–269.

Clifton, V. L. (2010). Sex and the human placenta: Mediating differential strategies of fetal growth and survival. *Placenta, 31*, S33–S39.

Coan, J. A., Beckes, L., & Allen, J. P. (2013). Childhood maternal support and social capital moderate the regulatory impact of social relationships in adulthood. *International Journal of Psychophysiology, 88*(3), 224–231.

Correia-Branco, A., Keating, E., & Martel, F. (2015). Maternal undernutrition and fetal developmental programming of obesity: The glucocorticoid connection. *Reproductive Sciences, 22*(2), 138–145.

Danese, A., & McEwen, B. S. (2012). Adverse childhood experiences, allostasis, allostatic load, and age-related disease. *Physiology and Behavior, 106*(1), 29–39.

Daskalakis, N. P., Bagot, R. C., Parker, K. J., Vinkers, C. H., & de Kloet, E. R. (2013). The three-hit concept of vulnerability and resilience: Towards understanding adaptation to early-life adversity outcome. *Psychoneuroendocrinology, 38*(9), 1858–1873.

De Kloet, E. R. (2014). From receptor balance to rational glucocorticoid therapy. *Endocrinology, 155*(8), 2754–2769.

Del Guidice, M., Ellis, B. J., & Shirtcliff, E. A. (2011). The adaptive calibration model of stress responsivity. *Neuroscience and Biobehavioral Reviews, 35*(7), 1562–1592.

DePasquale, C. E., Donzella, B., & Gunnar, M. R. (2019). Pubertal recalibration of cortisol reactivity following early life stress: A cross-sectional analysis. *Journal of Child Psychology and Psychiatry, and Allied Disciplines, 60*(5), 566–575.

Dettenborn, L., Tietze, A., Kirschbaum, C., & Stalder, T. (2012). The assessment of cortisol in human hair: Associations with sociodemographic variables and potential confounders. *Stress, 15*(6), 578–588.

Dettling, A. C., Gunnar, M. R., & Donzella, B. (1999). Cortisol levels of young children in full-day childcare centers: Relations with age and temperament. *Psychoneuroendocrinology, 24*(5), 519–536.

Fearon, R. M., Tomlinson, M., Kumsta, R., Skeen, S., Murray, L., Cooper, P. J., et al. (2017). Poverty, early care, and stress reactivity in adolescence: Findings from a prospective, longitudinal study in South Africa. *Development and Psychopathology, 29*(2), 449–464.

Feldman, R. (2015). Sensitive periods in human social development: New insights from research on oxytocin, synchrony, and high-risk parenting. *Development and Psychopathology, 27*(2), 369–395.

Feldman, R. (2017). The neurobiology of human attachments. *Trends in Cognitive Sciences, 21*(2), 80–99.

Fisher, P. A., Kim, H. K., Bruce, J., & Pears, K. C. (2012). Cumulative effects of prenatal substance exposure and early adversity on foster children's HPA axis reactivity during a psychosocial stressor. *International Journal of Behavioral Development, 36*(1), 29–35.

Gee, D. G., & Casey, B. J. (2015). The impact of developmental timing for stress and recovery. *Neurobiology of Stress, 1*, 184–194.

Gee, D. G., Gabard-Durnam, L., Telzer, E. H., Humphreys, K. L., Goff, B., Shapiro, M., et al. (2014). Maternal buffering of human amygdala–prefrontal circuitry during childhood but not adolescence. *Psychological Science, 25*(11), 2067–2078.

Gunnar, M. R. (2017). Social buffering of stress in development: A career perspective. *Perspectives on Psychological Science, 12*(3), 355–373.

Gunnar, M. R., & Davis, E. P. (2013). The effects of stress on early brain and behavioral development. In J. L. R. Rubenstein & P. Rakic (Eds.), *Comprehensive developmental neuroscience: Neural circuit development and function in the heathy and diseased brain* (pp. 447–465). Amsterdam: Academic Press.

Gunnar, M. R., & Donzella, B. (2002). Social regulation of the cortisol levels in early human development. *Psychoneuroendocrinology*, *27*(1–2), 199–220.

Gunnar, M. R., Kryzer, E., Van Ryzin, M. J., & Phillips, D. A. (2011). The import of the cortisol rise in child care differs as a function of behavioral inhibition. *Developmental Psychology*, *47*(3), 792–803.

Gunnar, M. R., Larson, M. C., Hertsgaard, L., Harris, M. L., & Brodersen, L. (1992). The stressfulness of separation among nine-month-old infants: Effects of social context variables and infant temperament. *Child Development*, *63*(2), 290–303.

Gunnar, M. R., Morison, S. J., Chisholm, K., & Schuder, M. (2001). Salivary cortisol levels in children adopted from Romanian orphanages. *Development and Psychopathology*, *13*(3), 611–628.

Gunnar, M. R., & Quevedo, K. (2007). The neurobiology of stress and development. *Annual Review of Psychology*, *58*(1), 145–173.

Gunnar, M. R., Sebanc, A. M., Tout, K., Donzella, B., & Van Dulmen, M. M. H. (2003). Peer rejection, temperament, and cortisol activity in preschoolers. *Developmental Psychobiology*, *43*(4), 346–358.

Gunnar, M. R., Talge, N. M., & Herrera, A. (2009). Stressor paradigms in developmental studies: What does and does not work to produce mean increases in salivary cortisol. *Psychoneuroendocrinology*, *34*(7), 953–967.

Gunnar, M. R., & Vazquez, D. (2006). Stress neurobiology and developmental psychopathology. *Development and Psychopathology*, *2*, 533–577.

Halligan, S. L., Herbert, J., Goodyer, I. M., & Murray, L. (2004). Exposure to postnatal depression predicts elevated cortisol in adolescent offspring. *Biological Psychiatry*, *55*(4), 376–381.

Hane, A. A., & Fox, N. A. (2006). Ordinary variations in maternal caregiving influence human infants' stress reactivity. *Psychological Science*, *17*(6), 550–556.

Hanson, J. L., Nacewicz, B. M., Sutterer, M. J., Cayo, A. A., Schaefer, S. M., Rudolph, K. D., et al. (2015). Behavior problems after early life stress: Contributions of the hippocampus and amygdala. *Biological Psychiatry*, *77*(4), 314–323.

Hare, T. A., Tottenham, N., Galvan, A., Voss, H. U., Glover, G. H., & Casey, B. J. (2008). Biological substrates of emotional reactivity and regulation in adolescence during an emotional go-nogo task. *Biological Psychiatry*, *63*(19), 927–934.

Hennessy, M. B., Kaiser, S., & Sachser, N. (2009). Social buffering of the stress response: Diversity, mechanisms, and functions. *Frontiers in Neuroendocrinology*, *30*(4), 470–482.

Herman, J. P. (2018). Regulation of hypothalamo-pituitary-adrenocortical responses to stressors by the nucleus of the solitary tract/dorsal vagal complex. *Cellular and Molecular Neurobiology*, *38*(1), 25–35.

Herman, J. P., Mcklveen, J. M., Ghosal, S., Kopp, B., Wulsin, A., Makinson, R., et al. (2016). Regulation of the hypothalamic-pituitary-adrenocortical stress response. *Comprehensive Physiology*, *6*(2), 603–621.

Hill, M. N., & Tasker, J. G. (2012). Endocannabinoid signaling, glucocorticoid-mediated negative feedback and regulation of the HPA axis. *Neuroscience*, *204*, 5–16.

Hostinar, C. E., Johnson, A. E., & Gunnar, M. R. (2015). Early social deprivation and the social buffering of cortisol stress responses in late childhood: An experimental study. *Developmental Psychology*, *51*(11), 1597–1608.

Hostinar, C. E., Sullivan, R. M., & Gunnar, M. R. (2014). Psychobiological mechanisms underlying the social buffering of the hypothalamic-pituitary-adrenocortical axis: A review of animal models and human studies across development. *Psychological Bulletin*, *140*(1), 256–282.

Howell, B. R., Grand, A. P., Mccormack, K. M., Shi, Y., Laprarie, J. L., Maestripieri, D., et al. (2014). Early adverse experience increases emotional reactivity in juvenile rhesus macaques: Relation to amygdala volume. *Developmental Psychobiology*, *56*(8), 1735–1746.

Howell, B. R., McCormack, K. M., Grand, A. P., Sawyer, N. T., Zhang, X., Maestripieri, D., et al. (2013). Brain white matter microstructure alterations in adolescent rhesus monkeys exposed to early life stress: Associations with high cortisol during infancy. *Biology of Mood & Anxiety Disorders*, *3*(1), 1–14.

Ishimoto, H., & Jaffe, R. B. (2011). Development and function of the human fetal adrenal cortex: A key component in the feto-placental unit. *Endocrine Reviews*, *32*(3), 317–355.

Jabès, A., Lavenex, P. B., Amaral, D. G., & Lavenex, P. (2011). Postnatal development of the hippocampal formation: A stereological study in macaque monkeys. *Journal of Comparative Neurology*, *519*(6), 1051–1070.

Joëls, M., & Baram, T. Z. (2009). The neuro-symphony of stress. *Nature Reviews Neuroscience*, *10*(6), 459–466.

Joëls, M., Karst, H., & Sarabdjitsingh, R. A. (2018). The stressed brain of humans and rodents. *Acta Physiologica*, *223*(2), 1–10.

Johnson, A. B., Mliner, S. B., Depasquale, C. E., Troy, M., & Gunnar, M. R. (2018). Attachment security buffers the HPA axis of toddlers growing up in poverty or near poverty: Assessment during pediatric well-child exams with inoculations. *Psychoneuroendocrinology*, *95*, 120–127.

Jonetz-Mentzel, L., & Wiedemann, G. (1993). Establishment of reference ranges for cortisol in neonates, infants, children and adolescents. *European Journal of Clinical Chemistry and Clinical Biochemistry*, *31*, 525–529.

King, L. S., Colich, N. L., LeMoult, J., Humphreys, K., Ordaz, S. J., Price, A. N., et al. (2017). The impact of the severity of early life stress on diurnal cortisol: The role of puberty. *Psychoneuroendocrinology*, *77*, 68–74.

Kirschbaum, C., & Hellhammer, D. H. (1994). Salivary cortisol in psychoneuroendocrine research: Recent developments and applications. *Psychoneuroendocrinology*, *19*(4), 313–333.

Kirschbaum, C., Kudielka, B. M., Gaab, J., Schommer, N. C., & Hellhammer, D. H. (1999). Impact of gender, menstrual cycle phase, and oral contraceptives on the activity of the hypothalamus-pituitary-adrenal axis. *Psychosomatic Medicine*, *61*, 154–162.

Klimes-Dougan, B., Hastings, P. D., Granger, D. A., Usher, B. A., & Zahn-Waxler, C. (2001). Adrenocortical activity in at-risk and normally developing adolescents: Individual differences in salivary cortisol basal levels, diurnal variation, and responses to social challenges. *Development and Psychopathology*, *13*(3), 695–719.

Koss, K. J., & Gunnar, M. R. (2018). Annual research review: Early adversity, the hypothalamic-pituitary-adrenocortical axis, and child psychopathology. *Journal of Child Psychology and Psychiatry*, *4*(59), 327–346.

Kuhlman, K. R., Vargas, I., Geiss, E. G., & Dopez-Duran, N. L. (2015). Age of trauma onset and HPA axis dysregulation among trauma-exposed youth. *Journal of Traumatic Stress*, *28*(6), 572–579.

Kumsta, R., Schlotz, W., Golm, D., Moser, D., Kennedy, M., Knights, N., et al. (2017). HPA axis dysregulation in adult adoptees twenty years after severe institutional deprivation in childhood. *Psychoneuroendocrinology*, *86*, 196–202.

Larose, M. P., Ouellet-Morin, I., Vitaro, F., Geoffroy, M. C., Ahun, M., Tremblay, R. E., et al. (2019). Impact of a social skills program on children's stress: A cluster randomized trial. *Psychoneuroendocrinology*, *104*, 115–121.

Lashansky, G., Saenger, P., Fishman, K., Gautier, T., Mayes, D., Berg, G., et al. (1991). Normative data for adrenal steroidogenesis in a healthy pediatric population: Age-and sex-related changes after adrenocorticotropin stimulation. *The Journal of Clinical Endocrinology & Metabolism*, *73*(3), 674–686.

Lazarus, R. S., & Folkman, S. (1984). *Stress, appraisal, and coping*. New York: Springer Publishing.

Leneman, K. B., Donzella, B., Desjardins, C. D., Miller, B. S., & Gunnar, M. R. (2018). The slope of cortisol from awakening to 30 min post-wake in post-institutionalized children and early adolescents. *Psychoneuroendocrinology*, *96*, 93–99.

Leneman, K. B., & Gunnar, M. R. (2018). Developmental timing of stress effects on the brain. In K. L. Harkness & E. P. Hayden (Eds.), *The Oxford handbook of stress and mental health*. Oxford, UK: Oxford University Press.

Levine, S. (1957). Infantile experience and resistance to physiological stress. *Science*, *126*(3270), 405.

Lupien, S. J., Juster, R. P., Raymond, C., & Marin, M. F. (2018). The effects of chronic stress on the human brain: From neurotoxicity, to vulnerability, to opportunity. *Frontiers in Neuroendocrinology*, *49*, 91–105.

Lupien, S. J., McEwen, B. S., Gunnar, M. R., & Heim, C. (2009). Effects of stress throughout the lifespan on the brain, behaviour and cognition. *Nature Reviews Neuroscience*, *10*(6), 434–445.

McEwen, B. S. (2000). Definition and concepts of stress. In G. Fink (Ed.), Vol. 3. *Encyclopedia of stress* (pp. 508–509). San Diego: Academic Press.

McEwen, B. S., & Stellar, E. (1993). Stress and the individual. *Archives of Internal Medicine*, *153*, 2093–2101.

McLaughlin, K. A., & Sheridan, M. A. (2016). Beyond cumulative risk: A dimensional approach to childhood adversity. *Current Directions in Psychological Science*, *25*(4), 239–245.

Miller, G. E., Chen, E., & Parker, K. J. (2011). Psychological stress in childhood and susceptibility to the chronic diseases of aging: Moving towards a model of behavioral and biological mechanisms. *Psychological Bulletin*, *137*(6), 959–997.

Morgan, J. T., & Amaral, D. G. (2014). Comparative analysis of the dendritic organization of principal neurons in the lateral and central nuclei of the rhesus macaque and rat amygdala. *Journal of Comparative Neurology*, *522*(3), 689–716.

Mulkey, S. B., & du Plessis, A. J. (2019). Autonomic nervous system development and its impact on neuropsychiatric outcome. *Pediatric Research*, *85*(2), 120–126.

Nazzari, S., Fearon, P., Rice, F., Dottori, N., Ciceri, F., Molteni, M., et al. (2019). Beyond the HPA-axis: Exploring maternal prenatal influences on birth outcomes and stress reactivity. *Psychoneuroendocrinology*, *101*, 253–262.

Netherton, C., Goodyer, I., Tamplin, A., & Herbert, J. (2004). Salivary cortisol and dehydroepiandrosterone in relation to puberty and gender. *Psychoneuroendocrinology*, *29*(2), 125–140.

O'Donnell, K. J., & Meaney, M. J. (2017). Fetal origins of mental health: The developmental origins of health and disease hypothesis. *American Journal of Psychiatry*, *174*(4), 319–328.

Oitzl, M. S., Champagne, D. L., van der Veen, R., & de Kloet, E. R. (2010). Brain development under stress: Hypotheses of glucocorticoid actions revisited. *Neuroscience and Biobehavioral Reviews*, *34*(6), 853–866.

Parker, K. J., Buckmaster, C. L., Schatzberg, A. F., & Lyons, D. M. (2004). Prospective investigation of stress inoculation in young monkeys. *Archives of General Psychiatry*, *61*(9), 933–941.

Parker, K. J., & Maestripieri, D. (2011). Identifying key features of early stressful experiences that produce stress vulnerability and resilience in primates. *Neuroscience and Biobehavioral Reviews*, *35*(7), 1466–1483.

Pearson, J., Tarabulsy, G. M., & Bussières, E. L. (2015). Foetal programming and cortisol secretion in early childhood: A meta-analysis of different programming variables. *Infant Behavior and Development*, *40*, 204–215.

Plotsky, P. M., & Meaney, M. J. (1993). Early, postnatal experience alters hypothalamic corticotropin-releasing factor (CRF) mRNA, median eminence CRF content and stress-induced release in adult rats. *Molecular Brain Research*, *18*(3), 195–200.

Porges, S. W., & Furman, S. A. (2011). The early development of the autonomic nervous system provides a neural platform for social behaviour: A polyvagal perspective. *Infant and Child Development, 20*, 106–118.

Quinn, T., Greaves, R., Badoer, E., & Walker, D. (2018). DHEA in prenatal and postnatal life: Implications for brain and behavior. *Vitamins and Hormones, 108*, 145–174. Elsevier Inc.

Reuben, A., Moffitt, T. E., Caspi, A., Belsky, D. W., Harrington, H., Schroeder, F., et al. (2016). Lest we forget: Comparing retrospective and prospective assessments of adverse childhood experiences in the prediction of adult health. *Journal of Child Psychology and Psychiatry, and Allied Disciplines, 57*(10), 1103–1112.

Romeo, R. D. (2010). Pubertal maturation and programming of hypothalamic-pituitary-adrenal reactivity. *Frontiers in Neuroendocrinology, 31*(2), 232–240.

Romeo, R. D. (2017). The impact of stress on the structure of the adolescent brain: Implications for adolescent mental health. *Brain Research, 1654*, 185–191.

Rosenfeld, P., van Eekelen, J. A. M., Levine, S., & de Kloet, E. R. (1993). Ontogeny of corticosteroid receptors in the brain. *Cellular and Molecular Neurobiology, 13*(4), 295–319.

Sánchez, M. M., Ladd, C. O., & Plotsky, P. M. (2001). Early adverse experience as a developmental risk factor for later psychopathology: Evidence from rodent and primate models. *Development and Psychopathology, 13*(3), 419–449.

Sandman, C. A., Glynn, L. M., & Davis, E. P. (2013). Is there a viability-vulnerability tradeoff? Sex differences in fetal programming. *Journal of Psychosomatic Research, 75*(4), 327–335.

Santiago, A., Aoki, C., & Sullivan, R. M. (2017). From attachment to independence: Stress hormone control of ecologically relevant emergence of infants' responses to threat. *Current Opinion in Behavioral Sciences, 14*, 78–85.

Sapolsky, R. M., Romero, L. M., & Munck, A. U. (2000). How do glucocorticoids influence stress responses? Integrating permissive, suppressive, stimulatory, and preparative actions. *Endocrine Reviews, 21*(1), 55–89.

Seltzer, L. J., Ziegler, T. E., & Pollak, S. D. (2010). Social vocalizations can release oxytocin in humans. *Proceedings of the Royal Society B: Biological Sciences, 277*(1694), 2661–2666.

Shirtcliff, E. A., Dismukes, A. R., Marceau, K., Ruttle, P. L., Simmons, J. G., & Han, G. (2015). A dual-axis approach to understanding neuroendocrine development. *Developmental Psychobiology, 57*(6), 643–653.

Sisk, C. L., & Zehr, J. L. (2005). Pubertal hormones organize the adolescent brain and behavior. *Frontiers in Neuroendocrinology, 26*(3–4), 163–174.

Sousa, N., & Almeida, O. F. X. (2002). Corticosteroids: Sculptors of the hippocampal formation. *Reviews in the Neurosciences, 13*, 59–84.

Sumner, M. M., Bernard, K., & Dozier, M. (2010). Young children's full-day patterns of cortisol production on child care days. *Archives of Pediatrics and Adolescent Medicine, 164*(6), 567–571.

Sumter, S. R., Bokhorst, C. L., Miers, A. C., Van Pelt, J., & Westenberg, P. M. (2010). Age and puberty differences in stress responses during a public speaking task: Do adolescents grow more sensitive to social evaluation? *Psychoneuroendocrinology, 35*(10), 1510–1516.

Tottenham, N., & Galván, A. (2016). Stress and the adolescent brain: Amygdala-prefrontal cortex circuitry and ventral striatum as developmental targets. *Neuroscience and Biobehavioral Reviews, 70*, 217–227.

Tottenham, N., Shapiro, M., Flannery, J., Caldera, C., & Sullivan, R. M. (2019). Parental presence switches avoidance to attraction learning in children. *Nature Human Behaviour, 3*, 1070–1077.

Tottenham, N., & Sheridan, M. A. (2010). A review of adversity, the amygdala and the hippocampus: A consideration of developmental timing. *Frontiers in Human Neuroscience, 3*, 1–18.

Uno, H., Lohmiller, L., Thieme, C., Kemnitz, J. W., Engle, M. J., Roecker, E. B., et al. (1990). Brain damage induced by prenatal exposure to dexamethasone in fetal rhesus macaques. I. Hippocampus. *Developmental Brain Research*, *53*(2), 157–167.

van Ijzendoorn, M. H., & Juffer, F. (2006). The Emanuel Miller Memorial Lecture 2006: Adoption as intervention. Meta-analytic evidence for massive catch-up and plasticity in physical, socio-emotional, and cognitive development. *Journal of Child Psychology and Psychiatry, and Allied Disciplines*, *47*(12), 1228–1245.

van Ijzendoorn, M. H., Palacios, J., Sonuga-Barke, E. J. S., Gunnar, M. R., Vorria, P., McCall, R. B., et al. (2011). I. Children in institutional care: Delayed development and resilience. *Monographs of the Society for Research in Child Development*, *76*(4), 8–30.

Vermeer, H. J., & van IJzendoorn, M. H. (2006). Children's elevated cortisol levels at daycare: A review and meta-analysis. *Early Childhood Research Quarterly*, *21*(3), 390–401.

Zhang, T. Y., Labonté, B., Wen, X. L., Turecki, G., & Meaney, M. J. (2013). Epigenetic mechanisms for the early environmental regulation of hippocampal glucocorticoid receptor gene expression in rodents and humans. *Neuropsychopharmacology*, *38*(1), 111–123.

Further reading

Porges, S. W. (1992). Vagal tone: A physiologic marker of stress vulnerability. *Pediatrics*, *90*(3), 498–504.

Pryce, C. R. (2008). Postnatal ontogeny of expression of the corticosteroid receptor genes in mammalian brains: Inter-species and intra-species differences. *Brain Research Reviews*, *57*(2), 596–605.

CHAPTER FOUR

The impact of childhood poverty on brain health: Emerging evidence from neuroimaging across the lifespan

Alexander J. Dufford[a,*], Pilyoung Kim[a], Gary W. Evans[b]
[a]Department of Psychology, University of Denver, Denver, CO, United States
[b]Departments of Design and Environmental Analysis and of Human Development, Cornell University, Ithaca, NY, United States
*Corresponding author: e-mail address: alexander.dufford@du.edu

Contents

Abstract

Experiencing poverty in childhood has been associated with increased risk for physical and mental health difficulties later in life. An emerging body of evidence suggests that brain development may be one mediator of this relation. In this chapter, we discuss evidence for an association between childhood poverty and brain structure/function. First, we examine the association from a lifespan perspective discussing studies at multiple developmental stages from the prenatal period to late adulthood. Second, we

International Review of Neurobiology, Volume 150
ISSN 0074-7742
https://doi.org/10.1016/bs.irn.2019.12.001

examine existing studies that link childhood poverty, brain development, and physical and mental health outcomes. Third, we discuss studies linking childhood poverty and environmental risks and protective factors. Lastly, we discuss suggestions for future studies including advances in network neuroscience, population neuroscience, using multiple imaging modalities, and the use of longitudinal neuroimaging studies. Overall, associations between childhood poverty, brain development, and development over the life course may help to both better understand and eventually reveal salient intervention strategies to mitigate social disparities in health.

Abbreviations

BOLD blood oxygenation level dependent
DTI diffusion tensor imaging
EEG electroencephalography
ERP event related potentials
fMRI functional magnetic resonance image
MRI magnetic resonance imaging
MT magnetization transfer
SE socioeconomic status

1. Introduction

There is an extensive literature linking childhood socioeconomic status (SES) to behavioral outcomes across several domains including physical health and mental health (Adler & Rehkopf, 2008; Kim, Evans, Chen, Miller, & Seeman, 2018; Wadsworth, Evans, Grant, Carter, & Duffy, 2016). SES refers to an individual or family's level of occupation, income, education or a combination of these indicators (McLoyd, 1998). Low SES is referred to as "poverty" when an individual or family's SES is below a certain threshold, typically the poverty threshold in which a country deems adequate based upon the family's income and number of people living in the home. For the chapter, SES will refer to when a combination of SES indicators is used (family income, education, occupation). When an indicator of SES is used in isolation, it will be specified such as the study used "family income" as a measure of SES.

A growing body of literature suggests potential mechanisms in which childhood poverty "gets under the skin" (Hackman, Farah, & Meaney, 2010; Kim et al., 2018; McEwen, 2012). The relations between childhood poverty and brain health has emerged as a potential pathway wherein childhood poverty may impact neural development, which in turn is associated with alterations in physical and mental outcomes (Farah, 2018; Hackman et al., 2010). Studies are beginning to expand to include examinations of the relations between childhood poverty and brain health in multiple periods of development (Farah, 2018; Johnson, Riis, & Noble, 2016). Further, studies are emerging that expand beyond the relation between childhood poverty and brain structure as variations of brain structure provide limited insight into brain function. Several reviews of the literature of the association between childhood poverty and brain exist but focus on specific developmental periods (Buckley, Broadley, & Cascio, 2019; Hackman et al., 2010).

Thus, this chapter adopts a lifespan perspective to examine the evidence about childhood poverty and brain development by focusing on the prenatal period and infancy, childhood, adolescence, and adulthood/late adulthood. Second, we discuss studies that have linked childhood poverty and brain development to physical health, behavioral outcomes, and mental health. Third, we discuss studies of both risk and protective factors related to childhood poverty and brain development. Lastly, we provide some ideas to enrich our understanding of childhood poverty and brain development using network neuroscience, population neuroscience, utilization of multiple neuroimaging modality, longitudinal, and mediation analysis studies. Studies included in the chapter were identified using keywords such as "poverty," "childhood poverty," "brain development," "brain structure," "brain function." The chapter is not a comprehensive review, rather a few studies we chosen for each section to illustrate a concept. Studies were not included that did not measure childhood SES or poverty or if brain development (structure/function) were not measured using a neuroimaging method such as structural MRI, fMRI, DTI, or EEG (see "Abbreviations" section for abbreviations used in the chapter).

2. Lifespan perspective

The early studies of the association between childhood poverty and brain development, borrowing from animal studies, focused on brain structure, specifically the hippocampus and amygdala. Evidence from animal models suggests that exposure to early adversity, such as high levels of stress, has causal effects on brain development (McEwen, 2008, 2012; Palma-Gudiel, Córdova-Palomera, Leza, & Fananás, 2015). However, there is extensive evidence that certain brain regions have heightened susceptibility to the effects of early adversity. Early life stress negatively affects neurogenesis in the hippocampus (Lajud & Torner, 2015), a brain region involved in learning and memory (Oomen et al., 2010). Further, early life stress is associated with increased dendritic arborization in the amygdala (Vyas, Mitra, Rao, & Chattarji, 2002), a region involved in salience processing, social behaviors, and fear learning (Santos, Mier, Kirsch, & Meyer-Lindenberg, 2011; van Marle, Hermans, Qin, & Fernández, 2009). This evidence suggests that similar processes may be operating in human children experiencing highly stressful situations such as poverty.

Advances in neuroimaging techniques, such as structural MRI, made it possible to examine gray matter volume of certain brain regions noninvasively in humans. While operational definitions of poverty varied (family income, parental education, income-to-needs ratio), studies consistently found a positive association between childhood SES and hippocampal volume in humans (Dufford, Bianco, & Kim, 2019; Hanson, Chandra, Wolfe, & Pollak, 2011; Jednoróg et al., 2012; Luby et al., 2013; Noble et al., 2015; Noble, Houston, Kan, & Sowell, 2012). There is less consistent evidence concerning amygdala structure in which some studies found a positive associations (Luby et al., 2013) and some studies found a negative association (Dufford et al., 2019). These early studies provided a critical foundation for the study of the relations between childhood poverty and brain development. However, these foundational findings do not reflect the enormous complexity of brain development over the life course, in particular different maturational periods of brain plasticity to environmental influences (Blair & Raver, 2012; Boyce, 2016; Tomalski & Johnson, 2010). This suggests that as the brain is changing across the lifespan, its relations with childhood poverty may also vary depending on developmental periods (Brito & Noble, 2014; Tottenham & Sheridan, 2010).

3. Prenatal period and infancy

The fetal brain undergoes rapid and unparalleled development and organization in utero (Stiles & Jernigan, 2010). As developmental periods of rapid development are accompanied with increased vulnerability for the brain (Johnson, 2005; Rodier, 1994; Westermann et al., 2007), it is critical to examine the potential relations between prenatal poverty and brain development. Infancy (particularly the first 2 years of life) is also a developmental period that is a critical foundation for brain maturation and subsequent cognitive and socioemotional development

(Knickmeyer et al., 2008; Uzgiris, 1973). However, the prenatal period and infancy is perhaps the least well understood time period concerning the relation between poverty and brain development. This could be due to the difficulty inherent in examining brain development during early life. While neuroimaging of children in this time period is becoming more accessible and feasible (Graham et al., 2015; Li et al., 2019), studies of the relation between prenatal experiences of poverty and brain development are still extremely limited. For example, studies using fetal MRI are quite limited and have not yet examined prenatal experiences of poverty with concurrent brain structure/function.

Studies of prenatal experiences typically involve assessments of brain development shortly after birth, to mitigate possible postnatal influences. Although not without difficulty, neonatal brain structure and function can be measured in vivo using MRI, typically during the neonate's natural sleep. One study found that low SES neonates have greater local brain volumes in the inferior frontal, cingulate, middle frontal and temporal pole, superior and middle occipital gyri (Spann, Bansal, Hao, Rosen, & Peterson, 2019). These findings contrast with the typical finding of a positive association between SES and brain volume. For example, using structural MRI, a study of 1-month-old infants found that lower SES was associated with lower cortical gray and deep gray matter volumes (Betancourt et al., 2016). Another study in infancy, using longitudinal neuroimaging, found that infants from lower income households had smaller or less gray matter volume in both the frontal and parietal lobes (Hanson et al., 2013). Infants from lower income families also had slower trajectories of growth in infancy and childhood. Using structural MRI and a large sample of infants, maternal education was positively associated with infant total white matter and gray matter volume (Knickmeyer et al., 2016). Whether SES is positively or negatively associated with infant brain structure require further examinations. Spann et al. (2019) suggest that early brain development involves both progressive and regressive processes. The first year of life is associated with dendritic arborization and glial cell multiplication to support synaptogenesis (Huttenlocher, De, Garey, & der Loos Van, 1982) while later in infancy and into middle childhood, growth is slowed by apoptosis and synaptic pruning (Huttenlocher, 1984). Due to the cross-sectional nature of existing studies, further longitudinal studies will be needed to determine how the direction of the association between SES and brain structure may change depending on which period of development is being examined. These findings highlight the complexity of this relation as well as how considering developmental trajectories is critical.

The association between childhood poverty and brain function has also been examined in infancy using EEG. EEG is a measure of brain function that measures electric brain activity through the scalp (Blinowska & Durka, 2006) and is widely used in infant studies. EEG can measure brain responses to specific time-epochs (ERP) or when individuals are engaged in observations of general stimuli known as resting baseline EEG (Marshall & Fox, 2007). In a study of 6–9 month olds, using resting baseline EEG, infants from low income households had lower frontal gamma power, resting brain oscillations association with language and cognitive skills in toddlers as well as an indicator of selective attention (Tomalski et al., 2013). Family SES was also associated with EEG activity during an error-detection task among older infants (aged 16–18 months). Infants from families with lower SES had lower activation of error-detection related EEG signal in the executive attention network (Conejero, Guerra, Abundis-Gutiérrez, & Rueda, 2018). Brain function in infancy can also be measured using resting-state fMRI which measures the intrinsic functional architecture of the brain by examining correlations in BOLD signal. Functional brain network development was associated with SES in a sample of infants at 6 months of age (Gao et al., 2014). SES had a significant positive association with measures of functional network maturation in the sensorimotor network and default mode network. These studies suggest that associations between childhood SES and brain structure/function can be detected early in development.

4. Childhood

In childhood, a study found a positive association between income-to-needs ratio and left hippocampus volumes and right amygdala volumes in children ages 6–12 (Luby et al., 2013). The associations between income-to-needs ratio and brain volumes were mediated by caregiving quality as well as stressful life events exposure (Luby et al., 2013). Studies have found positive associations between childhood poverty and cortical thickness and surface area in childhood (Lawson, Duda, Avants, Wu, & Farah, 2013; Noble et al., 2015). The most robust associations with brain surface morphometry appear to occur in prefrontal, anterior cingulate cortex, and medial temporal lobe regions.

While the findings for a relation between childhood poverty and brain structure are quite robust, the associations between childhood poverty and brain function are more limited. Prefrontal cortex activation was decreased in children (8–12 years old) experiencing lower SES during a stimulus-response mapping task (Sheridan, Sarsour, Jutte, D'Esposito, & Boyce, 2012). Another study in children (8–12 years old) found that children experiencing lower SES had lower hippocampal activation during a declarative memory task (Sheridan, How, Araujo, Schamberg, & Nelson, 2013). Two studies have found links between childhood SES, task-related functional activity, and reading deficits in childhood (ages 5–9) (Noble, Wolmetz, Ochs, Farah, & McCandliss, 2006; Raizada, Richards, Meltzoff, & Kuhl, 2008).

5. Adolescence

There is also emerging evidence of the association between childhood poverty and brain development in adolescence. Adolescence is also a time period in which the brain is developing rapidly, and prefrontal regions are developing at a slower rate than subcortical affective systems. Ziegler et al. used an innovative imaging technique known as MT, a marker of myelination and found that childhood disadvantage was associated with lower global MT and lower intra-cortical MT increases in sensory-motor, cingulate, insular, prefrontal, and subcortical areas (Ziegler et al., 2019). This suggests that childhood poverty may impact the myelination of white matter tracts that is occurring during adolescence. Weissman et al. also focused on adolescence; however, this study used resting-state fMRI to examine how changes in family income across adolescence were associated with connectivity of the default mode network (Weissman, Conger, Robins, Hastings, & Guyer, 2018). The connectivity of the default mode network depended on the slope of change in their family income with increases in family income across adolescence associated with greater prefrontal connectivity. Overall, studies from childhood and adolescent yield evidence suggesting relations between childhood poverty and subcortical gray matter structure (particularly the hippocampus), associations with cortical thickness and surface area (primarily prefrontal regions), as well as indications of an association with both structural and functional connectivity.

6. Adulthood and late adulthood

A series of studies have examined the prospective association between childhood poverty and brain structure/function in early adulthood (age 24). These longitudinal studies used measurements of family income-to-needs ratio measured at ages 9, 13, 17, and 24, finding a prospective relation between childhood poverty and prefrontal activation during an emotion regulation task (Kim et al., 2013), hippocampal function and visuospatial memory (Duval et al., 2017), and default mode network functional connectivity (Sripada, Swain, Evans, Welsh, & Liberzon, 2014). Using a multimodal approach, a study suggests a prospective association between childhood poverty (family income-to-needs ratio) and brain response to emotional faces in adulthood (Javanbakht et al., 2015). Specifically, lower family income-to-needs-ratio was associated with greater amygdala activation to threatening faces as well as lower prefrontal cortex activation to threatening faces. This suggests that childhood poverty may play a role in socioemotional development, specifically in the processing of social threats (Javanbakht et al., 2015).

There is also evidence that childhood SES is associated with brain structure/function in middle and late adulthood. In a sample of 42 middle aged, neurologically healthy men, a study reported differences in structural brain networks in men from neighborhoods varying in socioeconomic deprivation (Krishnadas et al., 2013). Men from the most socioeconomically deprived neighborhoods had structural brain networks that were less modular (a measure of the extent to which brain subnetwork organize into modules or units that are discrete entities with functions that are separable from other modules). Typically, higher modularity is a measure associated with more efficient information processing. These findings suggest that experiencing a neighborhood with high socioeconomic deprivation is associated with brain network efficiency (Rubinov & Sporns, 2010). Using a multimodal approach, for middle early (35–49) and middle late (50–64) adulthood, lower SES individuals had lower resting-state network segregation (a measure of how efficiently a brain network is organized) and diminished mean cortical thickness (Chan et al., 2018). These differences were not observed in the older age group (65–89). These findings suggest that high SES may be a protective factor for age-related decline in the brain. A study of participants in late adulthood (age 64) found that their recollections of their SES (home conditions and parental occupation) at age 11 was negatively associated with their hippocampal volume in late adulthood (Staff et al., 2012). Overall, these studies suggest that associations with childhood poverty can persist into adulthood and potentially have prospective relations into late adulthood. As demonstrated by the Chan et al. study, it will be important to examine how SES may impact brain aging and cognitive decline.

The current studies have provided evidence for associations between childhood poverty and brain development at multiple developmental stages. However, future studies will be needed to identify age sensitive periods for exposure to poverty. This has posed an analytic challenge as it is difficult to disentangle periods of development that have greater "sensitivity" to poverty exposure versus the duration of poverty exposure from the prenatal period until the development period in which brain development is being examined. To address this issue, future studies may benefit from utilizing human and animal model comparison studies (Perry et al., 2019), advanced statistical methods borrowing from structural equation modeling (Kievit et al., 2018), examining the efficacy of interventions at different developmental periods (Brody et al., 2017). For the remainder of the chapter, we discuss suggestions for study designs and methodologies that may facilitate understanding of this complex relation.

7. Links to physical health, behavioral outcomes, and mental health

An extensive literature documents that childhood poverty is associated with less positive physical and mental health outcomes (Adler & Rehkopf, 2008; Bradley & Corwyn, 2002; Kim et al., 2018; Wadsworth et al., 2016). Childhood poverty is associated with health outcomes later in life such as heart disease, obesity, and various infectious diseases (Haan, Kaplan, & Camacho, 1987; Lee, Andrew, Gebremariam, Lumeng, & Lee, 2014; Tomatis, 1997). The chronic stress associated with childhood poverty may underlie physical health and brain development sequelae of early poverty. Greater community socioeconomic disadvantage was associated with reduced cortical tissue volume, cortical surface area, and cortical thickness (Gianaros et al., 2017). Cardiometabolic risk (measures of adiposity, blood pressure, glucose, insulin, and lipids) mediated this association. Further, flatter diurnal cortisol decline (a measure of dysregulation) also mediated this association. These findings suggest a critical role of physical health and stress physiology in the link between poverty and brain development. Further supporting the role of physical health as a potential mediator, adiposity, cigarette smoking, and levels of C-reactive protein mediated the association between SES and white matter fractional anisotropy in adults (Gianaros, Marsland, Sheu, Erickson, & Verstynen, 2012). These findings suggest that future studies should focus on both inflammatory pathways as well as the association between childhood poverty, brain development, and dysregulation of stress response systems (Miller et al., 2009).

While early studies primarily established links between childhood poverty and the brain, recent studies have endeavored to extend these pathways to include how childhood poverty may be associated with behavioral outcomes and how measures of brain development may be also associated with these outcomes. It is critical to examine how structural and functional variations associated with childhood poverty are related to behaviors to avoid engaging in reverse inference (Ellwood-Lowe, Sacchet, & Gotlib, 2016). The association between childhood poverty and brain development could be a potential mechanistic pathway underlying the relations between childhood poverty and neurocognitive development (Farah, 2018; Johnson et al., 2016). The neurocognitive domains most robustly associated with SES appear to be executive functioning and language (Farah et al., 2006; Noble, Norman, & Farah, 2005). Lower family income was associated with lower performance on four cognitive assessments: a flanker task (inhibition), working memory, vocabulary, and reading (Noble et al., 2015). Cortical surface area was also positive associated with performance on these assessments. Of interest, surface area mediated the association between family income and cognitive performance on the flanker task and working memory task. Thus, lower cortical surface area may play an important role in the link between family income and executive functioning.

Language is another domain in which children that have experienced childhood poverty have difficulties (Farah et al., 2006). Cortical thickness mediated the link between SES and language abilities in children ages 3–21 (Khundrakpam et al., 2019). Cognitive abilities (such as executive functioning) and language play an important role in academic achievement. Hair et al. found that regional gray matter volumes from children 1.5 times below the federal poverty line were 3–4% points lower than the developmental norm (Hair, Hanson, Wolfe, & Pollak, 2015). Low-income children scored 4–7 points lower on standardized tests which was mediated by diminished gray matter volumes (Hair et al., 2015). This group of studies suggest that income inequalities in cognitive development may be mediated, at least in part, by underlying brain impacts of disadvantage.

In addition to relations between childhood poverty and behavioral outcomes such as cognition, it is also important to investigate the role of brain structure and function in mental health sequelae of early disadvantage. Children that have experienced poverty have an increased risk of developing mental disorders, such as anxiety and depression, later in life (Bradley & Corwyn, 2002; Gilman, Kawachi, Fitzmaurice, & Buka, 2002; Wadsworth et al., 2016). Using data from the PING study ($n = 1196$), lower family income and parental education was associated with lower amygdala volumes in adolescents (13–21) but not at younger ages (3–12). For a subsample ($n = 327$), lower parental education (not family income) was associated with greater internalizing symptoms (Merz, Tottenham, & Noble, 2018). Interestingly, smaller amygdala volumes were associated with greater levels of internalizing symptoms. These findings highlight the potential roles of childhood poverty and amygdala volume in the development of symptoms of anxiety and depression. Using task-based fMRI, gene methylation that was associated with SES in participants age 11–19 years old was associated with greater activation of the amygdala during a task in which participants viewed threatening stimuli (Swartz, Hariri, & Williamson, 2017). Further, this study found that the increases in amygdala reactivity moderated the relation between a positive family history of depression and later depressive symptoms. These studies suggest that variations in brain structure/function related to childhood SES have further associations with mental health symptoms.

8. Risks and protective factors

8.1 Cumulative risk

Future studies of the association between childhood poverty and brain development will benefit from being driven by theoretical models. Recent neuroimaging studies have adopted models of early adversity from developmental psychology. Compared to other forms of early adversity, childhood poverty is associated with a multitude of risk factors which portend deleterious development above and beyond singular risks (Evans, Li, & Whipple, 2013; Sameroff, Seifer, Zax, & Barocas, 1987). Cumulative risk typically includes both physical and psychosocial stressors, both of which commonly occur for individuals experiencing childhood poverty.

For example, one conceptualization of cumulative risk includes physical risks (noise, crowding, suboptimal housing) and psychosocial risks (family turmoil, child separation from family, violence) which are factors that children experiencing childhood poverty are more likely to be exposed to (Evans, 2004; Evans et al., 2013).

Recent studies of childhood poverty have included cumulative risk as a potential mediator of the relation between childhood poverty and the brain. As discussed, one study found that cumulative risk exposure (averaged between ages 13 and 17) mediated the association between childhood income-to-needs ratio and prefrontal cortex activity during an emotion regulation task (Kim et al., 2013). Greater cumulative risk exposure in childhood has also been associated with lower white matter fractional anisotropy in white matter tracts that were also associated with family income such as the cingulum bundle and superior longitudinal fasciculus (Dufford & Kim, 2017). These studies provide converging evidence that for childhood poverty, the exposure to multiple risks may be a potential pathway in which the stress associated with cumulative risk is associated with variations in brain development.

More typically studies of childhood poverty and brain development focus on a singular risk factors such as maltreatment or neglect. One dimension of early adversity, deprivation, is the absence of expected inputs that may impact neural proliferation and pruning (McLaughlin, Sheridan, & Lambert, 2014). The other dimension for early adversity, threat, is associated with alterations in fear learning processes (McLaughlin et al., 2014). The model suggests that different early adversities can be placed upon an axis based upon high versus low threat and high versus low deprivation. For example, neglect typically involves high deprivation but low threat whereas physical abuse typically involves high threat and low deprivation. However, childhood poverty is more difficult to describe accurately in terms of these dimensions as it involves both deprivation and threat.

8.2 Parenting interventions

It is critical for the understanding of the association between childhood poverty and brain development to examine potential protective effects that ameliorate this relation as well as potential interventions to mitigate the potential impacts of poverty on brain development. While protective factors and behavioral interventions for poverty exposure are abundant in the developmental psychology literature, there are a dearth of studies examining potential protective factors and interventions for the relation between childhood poverty and brain development. A longitudinal study of African American youth experiencing rural poverty has tested supportive parenting as a potential protective factor for the association between poverty and brain development (Brody et al., 2017). This study found that the amount of years living in poverty from ages 11 to 18 were associated with lower hippocampal (also dentate gyrus and CA3 subfields) and amygdala volumes. Participants in this study were a part of randomized, controlled trial design in which children and their parents were randomly assigned to a supportive parenting intervention or a control condition. Participants of parents that had participated in the supportive parenting intervention did not have the association between number of years living in poverty and hippocampal/amygdalar gray matter volume at age 25. These findings provide neural evidence for supportive parenting as a potential protective factor for children experiencing poverty. It also supports the potential utility of psychosocial interventions for ameliorating the association between childhood poverty and gray matter structure (Brody et al., 2017).

Using data collected from this study, there was evidence for the supportive parenting intervention having protective effects for resting-state functional connectivity (Brody, Yu, Nusslock et al., 2019). The number of years spent living in poverty from ages 11 to 18 had a prospective negative association with resting-state functional connectivity in the central executive and emotion regulation networks. The central executive network is involved in cognitive control, working memory, and is comprised of the dorsolateral prefrontal cortex and posterior parietal cortex. The emotion regulation network is involved in top-down control of limbic circuitry and is comprised of the inferior gyrus, middle temporal gyrus, and precentral gyrus. At the age of 25, participants whose parents had participated in the supportive parenting intervention did not have the association between years lived in poverty and central executive and emotion regulation network functional connectivity. Both studies highlight supportive parenting as a potential protective factors and intervention target for future studies examining childhood poverty and brain development.

Currently, in the developmental psychology and the developmental neuroscience literature, parenting has been highlighted as a protective factors and intervention target for the associations between childhood poverty and developmental outcomes (Brody, Yu, Miller, Ehrlich, & Chen, 2019). However, parents experiencing poverty are often under an enormous amount of stress and working multiple jobs. While there is evidence for intervening on positive parenting, this should not be the only protective factor that is examined using a developmental neuroscience framework. Potential protective factors for neuroimaging studies to focus on in the future could be individual-level factors such as coping or self-esteem (Harrison, Loxton, & Somhlaba, 2019), family emotional climate (Miller & Chen, 2013), school-level factors such as school-based interventions, or neighborhood-level factors such investments in after-school activities (Sharma, Mustanski, Dick, Bolland, & Kertes, 2019). Overall, the extant studies of the association between childhood poverty and brain development suggest that this relation is not fixed or immutable. Therefore, investments in studies that examine the mechanisms as well as the protective factors/targets for intervention are critical moving forward. In this section, we reviewed the latest developments in neuroimaging research examining the role of childhood SES in brain develop. We also discussed important future directions for the field to further the understanding of the neural embedding of childhood SES for psychological and physical health.

9. Future directions

9.1 Network neuroscience

As we have discussed, it is critical for the study of childhood poverty and brain development to study the brain beyond singular regions. It is also critical to combine multiple modalities and eventually combine multiple levels of analysis. Adopted from basic neuroscience studies, developmental neuroimaging has begun to examine the brain as a complex network of interconnected regions (Cao, Huang, Peng, Dong, & He, 2016; Zuo et al., 2017). Borrowing methods from graph theory (Rubinov & Sporns, 2010), this field of research is beginning to understand the complex interactions between brain regions as well as how they develop over time (Cao et al., 2016). Network neuroscience has intersected with an approach to studying the brain known as "connectomics." Connectomics attempts to provide a description of the connections between all regions of the brain (Behrens & Sporns, 2012). Connectomes at the macroscale can be measured using neuroimaging techniques in which each brain region is conceptualized as a node and each connection between nodes is an edge (Rubinov & Sporns, 2010). These analyses typically result in an adjacency or "connectivity" matrix representing the "connections" between all the nodes. This is typically measured by "structural connections" in the form of the connectivity strength between regions measured with diffusion weighted MRI (Hagmann et al., 2010). For functional connectomes, connectivity matrices typically are comprised of the correlation in BOLD timeseries among nodes calculated from resting-state fMRI or task-based fMRI. While connectomic approaches are becoming quite popular, the field of developmental connectomics is quite new and has only recently been used to examine the association between childhood poverty and brain development (Cao et al., 2016).

Studies that we have discussed have used structural connectomics approaches as well as combinations of structural and functional connectomic approaches. In addition to structural connectomes calculated based upon diffusion MRI, they can be calculate based upon correlations in structural measures such as cortical thickness (Alexander-Bloch, Raznahan, Bullmore, & Giedd, 2013). Further global metrics of the organization of brain networks can provide insight into how brain networks are organized. These graph theoretic measures can quantify the degree to which networks are integrated or segregated. Brain network segregation measures the brain's ability for specialized processing to occur in densely interconnected brain regions (Rubinov & Sporns, 2010). Graph theoretic measures can also measure brain integration which quantifies the brain's ability to combine specialized information from distributed brain regions (Rubinov & Sporns, 2010). Network segregation was examined based upon resting-state networks and a study found that individuals experiencing lower SES had reduced network segregation in middle aged individuals (Chan et al., 2018). This evidence suggests that SES is associated with the brain's ability to organize into units for specialized processing. Another measures of network segregation, modularity, measures the degree to which a network can be subdivided into clearly delineated and nonoverlapping groups (Chan et al., 2018). Comparing structural connectome modularity in individuals from areas experiencing low versus high socioeconomic deprivation, individuals from areas of high socioeconomic deprivation had lower network modularity (Krishnadas et al., 2013). Further evidence of an association between SES and network segregation was found using a large sample of 1012 youth ages 8–22. Using resting-state functional connectomes, youth experiencing high neighborhood SES had lower levels of initial local segregation and had the largest increases over time in local segregation (Tooley et al., 2019). Further, neighborhood SES moderated the relation between age and local segregation such that youth experiencing higher neighborhood SES had faster increases local segregation over time. This study suggests that SES may be an important factor that may influence how brain networks organize across development.

Evidence from these early studies utilizing network neuroscience suggests an association between poverty and brain network segregation. Future studies are needed to expand our understanding of the intersection of poverty and connectomics. As demonstrated by the studies discussed, network neuroscience is an advantageous approach to study this link. First, network neuroscience is a useful framework for multimodal studies as similar analysis can be conducted independent of modality (Scholtens & van den Heuvel, 2018). This facilitates comparison of structural networks to functional networks or can combine them (Wang, Dai, Gong, Zhou, & He, 2015). Network neuroscience can provide mathematical descriptions of how brain regions relate to one another and therefore are useful for studying brain network development over time (Rubinov & Sporns, 2010). Lastly, network neuroscience provides the ability to measure brain networks at multiple levels of analysis. For example, network neuroscience can unify descriptions of connections at the level of synapses, the level of individual neuronal connections, up to descriptions of the connectivity between populations of neurons and brain regions (Scholtens & van den Heuvel, 2018). As neuroimaging techniques improve in their resolution, studies will begin to have the ability to examine these multiple levels of analyses and network neuroscience can combine data across these levels; this will be critical for understanding potential biological processes underlying the association between childhood poverty and brain development.

9.2 Population neuroscience

Along with the emerging interest in connectomics, neuroscience has also become interested in population neuroscience. Concerns related to adequate statistical power for neuroimaging studies (Cremers, Wager, & Yarkoni, 2017), have motivated larger sample sizes. Population neuroscience typically involves data from large-scale repositories from many sites. By combining data from many different participants and research sites, the hope is to be able to have the statistical power and amount of data to be able to examine research questions not obtainable by smaller scale studies (Paus, 2010). One of the first example of population neuroscience studies has been influential in the study of the association between childhood poverty and brain development. A study used structural MRI data from the Pediatric Imaging, Neurocognition, and Genetics study (PING) (Jernigan et al., 2016).

The PING dataset is comprised of 1493 children aged from 3 to 20 years old. In addition to neuroimaging data, the PING data included genetic and cognitive data. The large sample size of the PING dataset afforded the ability to examine nonlinear associations between family income and surface morphometry. As discussed, this study found a logarithmic association between family and surface area. However, this dataset includes only bins of family income data and is cross-sectional. Therefore, future population neuroscience studies of childhood poverty should collect SES information (maternal education and family income-to-needs ratio) across multiple timepoints of development as well as have multiple neuroimaging measures across time.

Recently, more datasets of large samples are becoming available. The Philadelphia Neurodevelopmental Cohort (PNC) is another large-scale neuroimaging study (9498 children from 8 to 21 years old) in which the association between poverty and brain development can be studied (Satterthwaite et al., 2014). Although it is primarily focused on substance abuse, the Adolescent Brain Cognitive Development (ABCD) study has the potential to be leveraged to study the relation between childhood poverty and brain development using longitudinal data (Casey et al., 2018). The ABCD study is collecting data from over 10,000 youth aged 8–10 with additional data collected through adolescence.

Overall, data from population neuroscience studies provide large amounts of data to examine associations that may not be able to be properly studies in smaller samples. For the study of the relation between childhood poverty and brain development, population neuroscience may be critical to support analyses that typically require large amounts of data such as network analyses. Also, it is difficult to conduct longitudinal studies, despite these being critical to understand brain development. Therefore, large-scale multi-site studies may be able to achieve large longitudinal samples by combining data across sites. However, with large-scale neuroimaging studies such as PING, PNC, and ABCD it may be difficult to obtain detailed information about SES from each participant. We suggest that future studies of the association between childhood poverty and brain development will benefit from a combination of data from large-scale population neuroscience studies that may have limited detail in the collection of SES and smaller scale studies that can provide more detailed assessments of SES. Further, prior large samples (Gianaros et al., 2008; Staff et al., 2012) have relied on retrospective reports of childhood poverty which may be difficult to recall accurately. Therefore, future population neuroscience studies should collect information regarding childhood SES during childhood and at multiple timepoints across development.

9.3 Utilizing multiple neuroimaging modalities

While most of the early studies of the association between childhood poverty and brain development focused on gray matter volume, recent studies have expanded to other and complimentary neuroimaging modalities (Uludağ & Roebroeck, 2014). Use of multiple modalities can provide comprehensive measurements of multiple aspects of brain development (Biessmann, Plis, Meinecke, Eichele, & Muller, 2011). Ultimately, the goal is to provide converging and complimentary information about SES and brain development in order to have a deeper understanding. Beyond gray matter volume, as obtained from structural MRI, complimentary measures of gray matter structure called cortical thickness and surface area can be obtained (Brito & Noble, 2014). While cortical thickness and surface area contribute to cortical volume, they can provide unique information potential underlying processes. Cortical thickness and surface area have distinct genetic and development patterns (Panizzon et al., 2009; Wierenga, Langen, Oranje, & Durston, 2014). Often studies will measure cortical thickness and surface area together (Noble et al., 2015). While measuring these aspects of brain structure is important, it is still unclear what specific biological processes they capture. For example, cortical thinning observed across childhood was thought to reflect synaptic pruning (Gogtay et al., 2004; Nie, Li, & Shen, 2013); however, a recent study provided evidence that this thinning may be due to myelination of white matter (Natu et al., 2019). Therefore, studies of the relation between childhood poverty and brain structure may benefit from examining gray matter and white matter structure.

Recent studies have examined brain structure beyond gray matter volume to include white matter structure. White matter undergoes a protracted developmental trajectory which does not reach its peak until early adulthood (Lebel et al., 2012; Lebel, Treit, & Beaulieu, 2019). This may make it particularly susceptible to the influences of childhood poverty. As white matter development is linked with variations cognitive (Nagy, Westerberg, & Klingberg, 2004), language (Wong, Chandrasekaran, Garibaldi, & Wong, 2011), and socioemotional development (Versace et al., 2015), it is another important brain pathway to examine. Using data from the PING study, Ursache et al. found that SES had a positive association with white matter structure in a large sample of children ages 3–21. White matter structure can be measured in terms of its organization (known as fractional anisotropy). Higher SES was association with higher fractional anisotropy in the right parahippocampal cingulum and right superior corticostriate tract (Ursache, Noble, Pediatric Imaging, Neurocognition and Genetics Study, 2016). Another study found converging evidence of a link between family income-to-needs ratio and fractional anisotropy in the cingulum bundle in children ages 8–10. This study also found this association in the uncinate fasciculus, superior longitudinal fasciculus, inferior longitudinal fasciculus, and corticospinal tracts (Dufford & Kim, 2017). These studies provide evidence that the association between childhood poverty and brain development may not be unique to gray matter (regions involved in the computational processes of the brain) but also evident in white matter (tracts involved in transmission of information among regions).

Regarding multimodal neuroimaging, we also suggest that our understanding can be advanced by combining studies of structure and function. While we have a general idea of the functional roles of regions of the brain that are associated with childhood poverty, such as the hippocampus, it is often unclear what specific functional processed may be involved. Therefore, studies that combine structural and functional measures may give critical insight into how structural-functional relations may be associated with childhood poverty. Unfortunately, many studies of childhood poverty examine these associations separately. A study utilized both measures of cortical thickness and resting-state functional networks to examine differences in network organization based upon SES in adulthood (Chan et al., 2018). This study demonstrates that examining multiple modalities in concert may provide a deeper understanding of SES-brain relations. As each neuroimaging modality has its own strengths and weakness, combining them can enhance future studies. For example, EEG has exceptional temporal resolution but poor spatial resolution, while fMRI may not be able to examine rapid processes in the brain but has exceptional spatial resolution. Therefore, combining modalities may provide unique insight into the biological processes underlying the association between childhood poverty and brain development.

9.4 Longitudinal studies and mediation analysis

As we have discussed, the relation between childhood poverty and brain structure may depend on which developmental period is being examined.

Therefore, our understanding of this association will be enhanced using longitudinal data. Cross-sectional studies are limited in their ability to provide context related to the possible neurobiological underpinnings of the association. For example, several studies have reported a positive association between family income and hippocampal volume (Dufford et al., 2019; Hanson et al., 2011; Jednoróg et al., 2012; Luby et al., 2013; Noble et al., 2015, 2012). However, it is critical for these studies to consider the developmental stage of the participant in the study (Tottenham & Sheridan, 2010), while some studies would interpret this findings as a reduction in volume due to "damage" or stress altering the neural structure of the hippocampus, an alternative hypothesis is that at this developmental stage (middle childhood) the hippocampus is undergoing synaptic pruning which would also result in a decreased hippocampal volume (Callaghan & Tottenham, 2016; Tottenham & Sheridan, 2010). The experience of poverty may impact brain developmental by accelerating maturation, i.e., the lower hippocampal volume may be an indication of accelerated maturation, a response of the brain to adapt to environmental adversity (Callaghan & Tottenham, 2016; VanTieghem & Tottenham, 2017). As brain development is complex, it is critical for studies to have multiple timepoints of neuroimaging data. Longitudinal neuroimaging data allows for brain developmental trajectories to be examined; as discussed, these brain developmental trajectories can not only help inform interpretations regarding underlying neural processes but also the trajectories can be predictive of behavioral outcomes (Hanson et al., 2013). The brain trajectory approach from longitudinal data can also be informative regarding when trajectories diverge between children experiencing poverty and those not. These divergences may indicate development periods that are critical for intervention.

Longitudinal data also affords the ability to examine prospective relations between childhood poverty and brain development. As a large body of research has indicated that experiencing poverty in childhood has long-lasting associations with mental and physical health in adulthood, it is critical to examine prospective associations between childhood poverty and brain development. While identifying prospective relations between childhood poverty and brain development can demonstrate the potential long-lasting effect associations, this can also advance our understanding of potential protective factors that can ameliorate the association. We may be able to harness longitudinal studies to identify factors for those individuals that experienced poverty in childhood but have not experienced mental and physical health issues in adulthood. However, as discussed, it is critical for longitudinal studies to have accurate measures of childhood SES as retrospective recall of childhood indicators of SES may have limited accuracy (Jivraj, Goodman, Ploubidis, & de Oliveira, 2017).

Longitudinal studies also afford the ability to test potential mediators of the associations between childhood SES and brain development. Testing mediation in cross-sectional data has been shown to be biased (Maxwell & Cole, 2007), therefore it is critical for studies testing potential mediators of the relations between childhood poverty and brain development to utilize longitudinal data. The identification of mediators has the potential to inform potential mechanisms underlying the associations between childhood poverty and brain development as well as provide targets for interventions that may be particularly efficacious. However, to date, studies of the mediators of the relation between childhood poverty and brain development are rare. Caregiver support was found to mediate the association between income-to-needs ratio and hippocampal volume in children ages 6–12 (Luby et al., 2013). These findings suggest that the quality of caregiver support may play an important role in the associations between childhood SES and brain structure. Further, this may suggest that an intervention on parenting quality could potentially mitigate this relation; evidence for this has recently been found (Brody et al., 2017). The identification of potential mediators of the relation between childhood poverty and brain development can also test potential underlying mechanisms. Although it is problematic to conclude that a mediator identifies a "mechanism," evidence of statistical mediation is necessary but not sufficient to conclude a pathway is mechanistic (Tryon, 2018). Therefore, identifying mediators can be an important first step toward understanding the underlying mechanism of the associations between childhood poverty and brain structure. Subsequent analyses of the mediators that are identified are needed to establish causality and why a statistical mediation is observed in terms of in terms of a series of causal steps (Tryon, 2018).

Ultimately, a critical goal for future studies will be to build predictive models (Stringer & Tommerdahl, 2015) to identify individuals that have experienced childhood poverty and are at the highest risk for mental or physical health issues later in life. Identifying prospective relations is a critical first step in building predictive models; however, to improve the accuracy and generalizability of predictive models it is critical to test the model in a set of data that was not used to train the model (Picard & Cook, 1984). Using methods adopted from machine learning, cross-validation techniques can be used to test the accuracy of predictive models in an unbiased manner (Taylor, Ankerst, & Andridge, 2008; Woo, Chang, Lindquist, & Wager, 2017). Longitudinal studies are labor-intensive and expensive (especially the cost of multiple neuroimaging sessions); however, they are critical to provide a deeper understanding of the association between childhood poverty and brain development as well as provide the opportunity to examine mediators of the association.

10. Conclusions

Childhood poverty and brain development is a dynamic and complex process. First, the associations between childhood poverty and brain development have been identified across the life span from the prenatal period to aging populations. However, as discussed, we suggest that the directionality between childhood poverty and brain structure/function often depends on the developmental period being examined. While there is evidence of long-lasting prospective relations between childhood poverty and brain structure/function, in certain developmental periods this coupling may be stronger or weaker during other developmental periods. We also conclude that the association between childhood poverty and brain development is a complex process. Studies have begun to address this complexity by utilizing longitudinal studies, multimodal neuroimaging, network neuroscience, population neuroscience, theoretical models, plus examining health/behavioral outcomes, as well as protective factors. Further, we suggest that future studies of the association between childhood poverty and brain development may benefit from some of or a combination of these methods. Studies of the association between childhood poverty and brain health have and will continue to provide critical information about potential mechanisms underlying the association between childhood poverty and physical and mental health issues later in life.

References

Adler, N. E., & Rehkopf, D. H. (2008). US disparities in health: Descriptions, causes, and mechanisms. *Annual Review of Public Health*, *29*, 235–252.

Alexander-Bloch, A., Raznahan, A., Bullmore, E., & Giedd, J. (2013). The convergence of maturational change and structural covariance in human cortical networks. *Journal of Neuroscience*, *33*(7), 2889–2899.

Behrens, T. E., & Sporns, O. (2012). Human connectomics. *Current Opinion in Neurobiology*, *22*(1), 144–153.

Betancourt, L. M., Avants, B., Farah, M. J., Brodsky, N. L., Wu, J., Ashtari, M., et al. (2016). Effect of socioeconomic status (SES) disparity on neural development in female African-American infants at age 1 month. *Developmental Science*, *19*(6), 947–956.

Biessmann, F., Plis, S., Meinecke, F. C., Eichele, T., & Muller, K.-R. (2011). Analysis of multimodal neuroimaging data. *IEEE Reviews in Biomedical Engineering*, *4*, 26–58.

Blair, C., & Raver, C. C. (2012). Child development in the context of adversity: Experiential canalization of brain and behavior. *American Psychologist*, *67*(4), 309.

Blinowska, K., & Durka, P. (2006). Electroencephalography (EEG). In *Wiley encyclopedia of biomedical engineering*. Wiley.

Boyce, W. T. (2016). Differential susceptibility of the developing brain to contextual adversity and stress. *Neuropsychopharmacology*, *41*(1), 142.

Bradley, R. H., & Corwyn, R. F. (2002). Socioeconomic status and child development. *Annual Review of Psychology*, *53*(1), 371–399.

Brito, N. H., & Noble, K. G. (2014). Socioeconomic status and structural brain development. *Frontiers in Neuroscience*, *8*, 276.

Brody, G. H., Gray, J. C., Yu, T., Barton, A. W., Beach, S. R., Galván, A., et al. (2017). Protective prevention effects on the association of poverty with brain development. *JAMA Pediatrics*, *171*(1), 46–52.

Brody, G. H., Yu, T., Miller, G. E., Ehrlich, K. B., & Chen, E. (2019). Preventive parenting intervention during childhood and young black adults' unhealthful behaviors: A randomized controlled trial. *Journal of Child Psychology and Psychiatry*, *60*(1), 63–71.

Brody, G. H., Yu, T., Nusslock, R., Barton, A. W., Miller, G. E., Chen, E., et al. (2019). The protective effects of supportive parenting on the relationship between adolescent poverty and resting-state functional brain connectivity during adulthood. *Psychological Science*, *30*, 1040–1049. https://doi.org/10.1177/0956797619847989.

Buckley, L., Broadley, M., & Cascio, C. N. (2019). Socio-economic status and the developing brain in adolescence: A systematic review. *Child Neuropsychology*, *25*(7), 859–884.

Callaghan, B. L., & Tottenham, N. (2016). The stress acceleration hypothesis: Effects of early-life adversity on emotion circuits and behavior. *Current Opinion in Behavioral Sciences*, *7*, 76–81.

Cao, M., Huang, H., Peng, Y., Dong, Q., & He, Y. (2016). Toward developmental connectomics of the human brain. *Frontiers in Neuroanatomy*, *10*, 25.

Casey, B., Cannonier, T., Conley, M. I., Cohen, A. O., Barch, D. M., Heitzeg, M. M., et al. (2018). The adolescent brain cognitive development (ABCD) study: Imaging acquisition across 21 sites. *Developmental Cognitive Neuroscience*, *32*, 43–54.

Chan, M. Y., Na, J., Agres, P. F., Savalia, N. K., Park, D. C., & Wig, G. S. (2018). Socioeconomic status moderates age-related differences in the brain's functional network organization and anatomy across the adult lifespan. *Proceedings of the National Academy of Sciences of the United States of America*, *115*(22), E5144–E5153.

Conejero, Á., Guerra, S., Abundis-Gutiérrez, A., & Rueda, M. R. (2018). Frontal theta activation associated with error detection in toddlers: Influence of familial socioeconomic status. *Developmental Science*, *21*(1) e12494.

Cremers, H. R., Wager, T. D., & Yarkoni, T. (2017). The relation between statistical power and inference in fMRI. *PLoS One*, *12*(11) e0184923.

Dufford, A. J., Bianco, H., & Kim, P. (2019). Socioeconomic disadvantage, brain morphometry, and attentional bias to threat in middle childhood. *Cognitive, Affective, & Behavioral Neuroscience*, *19*(2), 309–326.

Dufford, A. J., & Kim, P. (2017). Family income, cumulative risk exposure, and white matter structure in middle childhood. *Frontiers in Human Neuroscience*, *11*, 547.

Duval, E. R., Garfinkel, S. N., Swain, J. E., Evans, G. W., Blackburn, E. K., Angstadt, M., et al. (2017). Childhood poverty is associated with altered hippocampal function and visuospatial memory in adulthood. *Developmental Cognitive Neuroscience*, *23*, 39–44.

Ellwood-Lowe, M. E., Sacchet, M. D., & Gotlib, I. H. (2016). The application of neuroimaging to social inequity and language disparity: A cautionary examination. *Developmental Cognitive Neuroscience*, *22*, 1–8.

Evans, G. W. (2004). The environment of childhood poverty. *American Psychologist, 59*(2), 77.
Evans, G. W., Li, D., & Whipple, S. S. (2013). Cumulative risk and child development. *Psychological Bulletin, 139*(6), 1342.
Farah, M. J. (2018). Socioeconomic status and the brain: Prospects for neuroscience-informed policy. *Nature Reviews Neuroscience, 19*(7), 428–438.
Farah, M. J., Shera, D. M., Savage, J. H., Betancourt, L., Giannetta, J. M., Brodsky, N. L., et al. (2006). Childhood poverty: Specific associations with neurocognitive development. *Brain Research, 1110*(1), 166–174.
Gao, W., Alcauter, S., Elton, A., Hernandez-Castillo, C. R., Smith, J. K., Ramirez, J., et al. (2014). Functional network development during the first year: Relative sequence and socioeconomic correlations. *Cerebral Cortex, 25*(9), 2919–2928.
Gianaros, P. J., Horenstein, J. A., Hariri, A. R., Sheu, L. K., Manuck, S. B., Matthews, K. A., et al. (2008). Potential neural embedding of parental social standing. *Social Cognitive and Affective Neuroscience, 3*(2), 91–96.
Gianaros, P. J., Kuan, D. C.-H., Marsland, A. L., Sheu, L. K., Hackman, D. A., Miller, K. G., et al. (2017). Community socioeconomic disadvantage in midlife relates to cortical morphology via neuroendocrine and cardiometabolic pathways. *Cerebral Cortex, 27*(1), 460–473.
Gianaros, P. J., Marsland, A. L., Sheu, L. K., Erickson, K. I., & Verstynen, T. D. (2012). Inflammatory pathways link socioeconomic inequalities to white matter architecture. *Cerebral Cortex, 23*(9), 2058–2071.
Gilman, S. E., Kawachi, I., Fitzmaurice, G. M., & Buka, S. L. (2002). Socioeconomic status in childhood and the lifetime risk of major depression. *International Journal of Epidemiology, 31*(2), 359–367.
Gogtay, N., Giedd, J. N., Lusk, L., Hayashi, K. M., Greenstein, D., Vaituzis, A. C., et al. (2004). Dynamic mapping of human cortical development during childhood through early adulthood. *Proceedings of the National Academy of Sciences of the United States of America, 101*(21), 8174–8179.
Graham, A. M., Pfeifer, J. H., Fisher, P. A., Lin, W., Gao, W., & Fair, D. A. (2015). The potential of infant fMRI research and the study of early life stress as a promising exemplar. *Developmental Cognitive Neuroscience, 12*, 12–39.
Haan, M., Kaplan, G. A., & Camacho, T. (1987). Poverty and health prospective evidence from the alameda county study. *American Journal of Epidemiology, 125*(6), 989–998.
Hackman, D. A., Farah, M. J., & Meaney, M. J. (2010). Socioeconomic status and the brain: Mechanistic insights from human and animal research. *Nature Reviews Neuroscience, 11*(9), 651.
Hagmann, P., Cammoun, L., Gigandet, X., Gerhard, S., Grant, P. E., Wedeen, V., et al. (2010). MR connectomics: Principles and challenges. *Journal of Neuroscience Methods, 194*(1), 34–45.
Hair, N. L., Hanson, J. L., Wolfe, B. L., & Pollak, S. D. (2015). Association of child poverty, brain development, and academic achievement. *JAMA Pediatrics, 169*(9), 822–829.
Hanson, J. L., Chandra, A., Wolfe, B. L., & Pollak, S. D. (2011). Association between income and the hippocampus. *PLoS One, 6*(5) e18712.
Hanson, J. L., Hair, N., Shen, D. G., Shi, F., Gilmore, J. H., Wolfe, B. L., et al. (2013). Family poverty affects the rate of human infant brain growth. *PLoS One, 8*(12) e80954.
Harrison, C., Loxton, H., & Somhlaba, N. Z. (2019). Stress and coping: Considering the influence of psychological strengths on the mental health of at-risk south African adolescents. *Child Care in Practice, 25*, 1–15.
Huttenlocher, P. R. (1984). Synapse elimination and plasticity in developing human cerebral cortex. *American Journal of Mental Deficiency, 88*, 488–496.
Huttenlocher, P. R., De, C. C., Garey, L. J., & der Loos Van, H. (1982). Synaptic development in human cerebral cortex. *International Journal of Neurology, 16*, 144–154.

Javanbakht, A., King, A. P., Evans, G. W., Swain, J. E., Angstadt, M., Phan, K. L., et al. (2015). Childhood poverty predicts adult amygdala and frontal activity and connectivity in response to emotional faces. *Frontiers in Behavioral Neuroscience*, *9*, 154.

Jednoróg, K., Altarelli, I., Monzalvo, K., Fluss, J., Dubois, J., Billard, C., et al. (2012). The influence of socioeconomic status on children's brain structure. *PLoS One*, 7(8) e42486.

Jernigan, T. L., Brown, T. T., Hagler, D. J., Jr., Akshoomoff, N., Bartsch, H., Newman, E., et al. (2016). The pediatric imaging, neurocognition, and genetics (PING) data repository. *NeuroImage*, *124*, 1149–1154.

Jivraj, S., Goodman, A., Ploubidis, G. B., & de Oliveira, C. (2017). Testing comparability between retrospective life history data and prospective birth cohort study data. *The Journals of Gerontology: Series B, Psychological Sciences and Social Sciences*, gbx042. https://doi.org/10.1093/geronb/gbx042.

Johnson, M. H. (2005). Sensitive periods in functional brain development: Problems and prospects. *Developmental Psychobiology*, *46*(3), 287–292.

Johnson, S. B., Riis, J. L., & Noble, K. G. (2016). State of the art review: Poverty and the developing brain. *Pediatrics*, *137*(4) e20153075.

Khundrakpam, B., Choudhury, S., Vainik, U., Al-Sharif, N., Bhutani, N., & Evans, A. (2019). Non-linear effects of socioeconomic status on brain development: associations between parental occupation, cortical thickness and language skills in childhood and adolescence. *bioRxiv*. https://doi.org/10.1101/575993.

Kievit, R. A., Brandmaier, A. M., Ziegler, G., Van Harmelen, A.-L., de Mooij, S. M., Moutoussis, M., et al. (2018). Developmental cognitive neuroscience using latent change score models: A tutorial and applications. *Developmental Cognitive Neuroscience*, *33*, 99–117.

Kim, P., Evans, G. W., Angstadt, M., Ho, S. S., Sripada, C. S., Swain, J. E., et al. (2013). Effects of childhood poverty and chronic stress on emotion regulatory brain function in adulthood. *Proceedings of the National Academy of Sciences of the United States of America*, *110*(46), 18442–18447.

Kim, P., Evans, G. W., Chen, E., Miller, G., & Seeman, T. (2018). How socioeconomic disadvantages get under the skin and into the brain to influence health development across the lifespan. In *Handbook of life course health development* (pp. 463–497). Cham: Springer.

Knickmeyer, R. C., Gouttard, S., Kang, C., Evans, D., Wilber, K., Smith, J. K., et al. (2008). A structural MRI study of human brain development from birth to 2 years. *Journal of Neuroscience*, *28*(47), 12176–12182.

Knickmeyer, R. C., Xia, K., Lu, Z., Ahn, M., Jha, S. C., Zou, F., et al. (2016). Impact of demographic and obstetric factors on infant brain volumes: A population neuroscience study. *Cerebral Cortex*, *27*(12), 5616–5625.

Krishnadas, R., McLean, J., Batty, G. D., Burns, H., Deans, K. A., Ford, I., et al. (2013). Socioeconomic deprivation and cortical morphology: Psychological, social, and biological determinants of ill health study. *Psychosomatic Medicine*, *75*(7), 616–623.

Lajud, N., & Torner, L. (2015). Early life stress and hippocampal neurogenesis in the neonate: Sexual dimorphism, long term consequences and possible mediators. *Frontiers in Molecular Neuroscience*, *8*, 3.

Lawson, G. M., Duda, J. T., Avants, B. B., Wu, J., & Farah, M. J. (2013). Associations between children's socioeconomic status and prefrontal cortical thickness. *Developmental Science*, *16*(5), 641–652.

Lebel, C., Gee, M., Camicioli, R., Wieler, M., Martin, W., & Beaulieu, C. (2012). Diffusion tensor imaging of white matter tract evolution over the lifespan. *NeuroImage*, *60*(1), 340–352.

Lebel, C., Treit, S., & Beaulieu, C. (2019). A review of diffusion MRI of typical white matter development from early childhood to young adulthood. *NMR in Biomedicine*, *32*(4) e3778.

Lee, H., Andrew, M., Gebremariam, A., Lumeng, J. C., & Lee, J. M. (2014). Longitudinal associations between poverty and obesity from birth through adolescence. *American Journal of Public Health, 104*(5), e70–e76.

Li, G., Wang, L., Yap, P.-T., Wang, F., Wu, Z., Meng, Y., et al. (2019). Computational neuroanatomy of baby brains: A review. *NeuroImage, 185*, 906–925.

Luby, J., Belden, A., Botteron, K., Marrus, N., Harms, M. P., Babb, C., et al. (2013). The effects of poverty on childhood brain development: The mediating effect of caregiving and stressful life events. *JAMA Pediatrics, 167*(12), 1135–1142.

Marshall, P. J., & Fox, N. A. (2007). Infant EEG and ERP in relation to social and emotional development. In *Infant EEG and event-related Potentials* (pp. 227–250). Hove, East Sussex: Psychology Press.

Maxwell, S. E., & Cole, D. A. (2007). Bias in cross-sectional analyses of longitudinal mediation. *Psychological Methods, 12*(1), 23.

McEwen, B. S. (2008). Understanding the potency of stressful early life experiences on brain and body function. *Metabolism, 57*, S11–S15.

McEwen, B. S. (2012). Brain on stress: How the social environment gets under the skin. *Proceedings of the National Academy of Sciences of the United States of America, 109*(Suppl. 2), 17180–17185.

McLaughlin, K. A., Sheridan, M. A., & Lambert, H. K. (2014). Childhood adversity and neural development: Deprivation and threat as distinct dimensions of early experience. *Neuroscience & Biobehavioral Reviews, 47*, 578–591.

McLoyd, V. C. (1998). Socioeconomic disadvantage and child development. *American Psychologist, 53*(2), 185.

Merz, E. C., Tottenham, N., & Noble, K. G. (2018). Socioeconomic status, amygdala volume, and internalizing symptoms in children and adolescents. *Journal of Clinical Child & Adolescent Psychology, 47*(2), 312–323.

Miller, G. E., & Chen, E. (2013). The biological residue of childhood poverty. *Child Development Perspectives*, 7(2), 67–73.

Miller, G. E., Chen, E., Fok, A. K., Walker, H., Lim, A., Nicholls, E. F., et al. (2009). Low early-life social class leaves a biological residue manifested by decreased glucocorticoid and increased proinflammatory signaling. *Proceedings of the National Academy of Sciences, 106*(34), 14716–14721.

Nagy, Z., Westerberg, H., & Klingberg, T. (2004). Maturation of white matter is associated with the development of cognitive functions during childhood. *Journal of Cognitive Neuroscience, 16*(7), 1227–1233.

Natu, V. S., Gomez, J., Barnett, M., Jeska, B., Kirilina, E., Jaeger, C., et al. (2019). Apparent thinning of human visual cortex during childhood is associated with myelination. *Proceedings of the National Academy of Sciences of the United States of America, 116*(41), 20750–20759.

Nie, J., Li, G., & Shen, D. (2013). Development of cortical anatomical properties from early childhood to early adulthood. *NeuroImage, 76*, 216–224.

Noble, K. G., Houston, S. M., Brito, N. H., Bartsch, H., Kan, E., Kuperman, J. M., et al. (2015). Family income, parental education and brain structure in children and adolescents. *Nature Neuroscience, 18*(5), 773.

Noble, K. G., Houston, S. M., Kan, E., & Sowell, E. R. (2012). Neural correlates of socioeconomic status in the developing human brain. *Developmental Science, 15*(4), 516–527.

Noble, K. G., Norman, M. F., & Farah, M. J. (2005). Neurocognitive correlates of socioeconomic status in kindergarten children. *Developmental Science, 8*(1), 74–87.

Noble, K. G., Wolmetz, M. E., Ochs, L. G., Farah, M. J., & McCandliss, B. D. (2006). Brain–behavior relationships in reading acquisition are modulated by socioeconomic factors. *Developmental Science, 9*(6), 642–654.

Oomen, C. A., Soeters, H., Audureau, N., Vermunt, L., Van Hasselt, F. N., Manders, E. M., et al. (2010). Severe early life stress hampers spatial learning and neurogenesis, but improves hippocampal synaptic plasticity and emotional learning under high-stress conditions in adulthood. *Journal of Neuroscience*, *30*(19), 6635–6645.

Palma-Gudiel, H., Córdova-Palomera, A., Leza, J. C., & Fananás, L. (2015). Glucocorticoid receptor gene (NR3C1) methylation processes as mediators of early adversity in stress-related disorders causality: A critical review. *Neuroscience & Biobehavioral Reviews*, *55*, 520–535.

Panizzon, M. S., Fennema-Notestine, C., Eyler, L. T., Jernigan, T. L., Prom-Wormley, E., Neale, M., et al. (2009). Distinct genetic influences on cortical surface area and cortical thickness. *Cerebral Cortex*, *19*(11), 2728–2735.

Paus, T. (2010). Population neuroscience: Why and how. *Human Brain Mapping*, *31*(6), 891–903.

Perry, R. E., Finegood, E. D., Braren, S. H., Dejoseph, M. L., Putrino, D. F., Wilson, D. A., et al. (2019). Developing a neurobehavioral animal model of poverty: Drawing cross-species connections between environments of scarcity-adversity, parenting quality, and infant outcome. *Development and Psychopathology*, *31*(2), 399–418.

Picard, R. R., & Cook, R. D. (1984). Cross-validation of regression models. *Journal of the American Statistical Association*, *79*(387), 575–583.

Raizada, R. D., Richards, T. L., Meltzoff, A., & Kuhl, P. K. (2008). Socioeconomic status predicts hemispheric specialisation of the left inferior frontal gyrus in young children. *NeuroImage*, *40*(3), 1392–1401.

Rodier, P. M. (1994). Vulnerable periods and processes during central nervous system development. *Environmental Health Perspectives*, *102*(Suppl. 2), 121–124.

Rubinov, M., & Sporns, O. (2010). Complex network measures of brain connectivity: Uses and interpretations. *NeuroImage*, *52*(3), 1059–1069.

Sameroff, A. J., Seifer, R., Zax, M., & Barocas, R. (1987). Early indicators of developmental risk: Rochester longitudinal study. *Schizophrenia Bulletin*, *13*(3), 383–394.

Santos, A., Mier, D., Kirsch, P., & Meyer-Lindenberg, A. (2011). Evidence for a general face salience signal in human amygdala. *NeuroImage*, *54*(4), 3111–3116.

Satterthwaite, T. D., Elliott, M. A., Ruparel, K., Loughead, J., Prabhakaran, K., Calkins, M. E., et al. (2014). Neuroimaging of the Philadelphia neurodevelopmental cohort. *NeuroImage*, *86*, 544–553.

Scholtens, L. H., & van den Heuvel, M. P. (2018). Multimodal connectomics in psychiatry: Bridging scales from micro to macro. *Biological Psychiatry: Cognitive Neuroscience and Neuroimaging*, *3*(9), 767–776.

Sharma, S., Mustanski, B., Dick, D., Bolland, J., & Kertes, D. A. (2019). Protective factors buffer life stress and behavioral health outcomes among high-risk youth. *Journal of Abnormal Child Psychology*, *47*, 1289–1301.

Sheridan, M. A., How, J., Araujo, M., Schamberg, M. A., & Nelson, C. A. (2013). What are the links between maternal social status, hippocampal function, and HPA axis function in children? *Developmental Science*, *16*(5), 665–675.

Sheridan, M. A., Sarsour, K., Jutte, D., D'Esposito, M., & Boyce, W. T. (2012). The impact of social disparity on prefrontal function in childhood. *PLoS One*, 7(4) e35744.

Spann, M. N., Bansal, R., Hao, X., Rosen, T. S., & Peterson, B. S. (2019). Prenatal socioeconomic status and social support are associated with neonatal brain morphology, toddler language and psychiatric symptoms. *Child Neuropsychology*, *6*, 1–19.

Sripada, R. K., Swain, J. E., Evans, G. W., Welsh, R. C., & Liberzon, I. (2014). Childhood poverty and stress reactivity are associated with aberrant functional connectivity in default mode network. *Neuropsychopharmacology*, *39*(9), 2244.

Staff, R. T., Murray, A. D., Ahearn, T. S., Mustafa, N., Fox, H. C., & Whalley, L. J. (2012). Childhood socioeconomic status and adult brain size: Childhood socioeconomic status influences adult hippocampal size. *Annals of Neurology*, *71*(5), 653–660.

Stiles, J., & Jernigan, T. L. (2010). The basics of brain development. *Neuropsychology Review, 20*(4), 327–348.
Stringer, S., & Tommerdahl, J. (2015). Building bridges between neuroscience, cognition and education with predictive modeling. *Mind, Brain, and Education, 9*(2), 121–126.
Swartz, J. R., Hariri, A. R., & Williamson, D. E. (2017). An epigenetic mechanism links socioeconomic status to changes in depression-related brain function in high-risk adolescents. *Molecular Psychiatry, 22*(2), 209.
Taylor, J. M., Ankerst, D. P., & Andridge, R. R. (2008). Validation of biomarker-based risk prediction models. *Clinical Cancer Research, 14*(19), 5977–5983.
Tomalski, P., & Johnson, M. H. (2010). The effects of early adversity on the adult and developing brain. *Current Opinion in Psychiatry, 23*(3), 233–238.
Tomalski, P., Moore, D. G., Ribeiro, H., Axelsson, E. L., Murphy, E., Karmiloff-Smith, A., et al. (2013). Socioeconomic status and functional brain development–associations in early infancy. *Developmental Science, 16*(5), 676–687.
Tomatis, L. (1997). Poverty and cancer. *Sociological Methods and Research, 160*(150), 140.
Tooley, U. A., Mackey, A. P., Ciric, R., Ruparel, K., Moore, T. M., Gur, R. C., et al. (2019). Associations between neighborhood SES and functional brain network development. *Cerebral Cortex*.
Tottenham, N., & Sheridan, M. A. (2010). A review of adversity, the amygdala and the hippocampus: A consideration of developmental timing. *Frontiers in Human Neuroscience, 3*, 68.
Tryon, W. W. (2018). Mediators and mechanisms. *Clinical Psychological Science, 6*(5), 619–628.
Uludağ, K., & Roebroeck, A. (2014). General overview on the merits of multimodal neuroimaging data fusion. *NeuroImage, 102*, 3–10.
Ursache, A., Noble, K. G., & Pediatric Imaging, Neurocognition and Genetics Study. (2016). Socioeconomic status, white matter, and executive function in children. *Brain and Behavior, 6*(10) e00531.
Uzgiris, I. C. (1973). Patterns of cognitive development in infancy. *Merrill-Palmer Quarterly of Behavior and Development, 19*(3), 181–204.
van Marle, H. J., Hermans, E. J., Qin, S., & Fernández, G. (2009). From specificity to sensitivity: How acute stress affects amygdala processing of biologically salient stimuli. *Biological Psychiatry, 66*(7), 649–655.
VanTieghem, M. R., & Tottenham, N. (2017). Neurobiological programming of early life stress: Functional development of amygdala-prefrontal circuitry and vulnerability for stress-related psychopathology. In *Behavioral neurobiology of PTSD* (pp. 117–136): Springer.
Versace, A., Acuff, H., Bertocci, M. A., Bebko, G., Almeida, J. R., Perlman, S. B., et al. (2015). White matter structure in youth with behavioral and emotional dysregulation disorders: A probabilistic tractographic study. *JAMA Psychiatry, 72*(4), 367–376.
Vyas, A., Mitra, R., Rao, B. S., & Chattarji, S. (2002). Chronic stress induces contrasting patterns of dendritic remodeling in hippocampal and amygdaloid neurons. *Journal of Neuroscience, 22*(15), 6810–6818.
Wadsworth, M. E., Evans, G. W., Grant, K., Carter, J. S., & Duffy, S. (2016). Poverty and the development of psychopathology. In D. Cicchetti (Ed.), *Developmental psychopathology* (pp. 136–179): John Wiley & Sons Inc.
Wang, Z., Dai, Z., Gong, G., Zhou, C., & He, Y. (2015). Understanding structural-functional relationships in the human brain: A large-scale network perspective. *The Neuroscientist, 21*(3), 290–305.
Weissman, D. G., Conger, R. D., Robins, R. W., Hastings, P. D., & Guyer, A. E. (2018). Income change alters default mode network connectivity for adolescents in poverty. *Developmental Cognitive Neuroscience, 30*, 93–99.

Westermann, G., Mareschal, D., Johnson, M. H., Sirois, S., Spratling, M. W., & Thomas, M. S. (2007). Neuroconstructivism. *Developmental Science*, *10*(1), 75–83.
Wierenga, L. M., Langen, M., Oranje, B., & Durston, S. (2014). Unique developmental trajectories of cortical thickness and surface area. *NeuroImage*, *87*, 120–126.
Wong, F. C., Chandrasekaran, B., Garibaldi, K., & Wong, P. C. (2011). White matter anisotropy in the ventral language pathway predicts sound-to-word learning success. *Journal of Neuroscience*, *31*(24), 8780–8785.
Woo, C.-W., Chang, L. J., Lindquist, M. A., & Wager, T. D. (2017). Building better biomarkers: Brain models in translational neuroimaging. *Nature Neuroscience*, *20*(3), 365.
Ziegler, G., Moutoussis, M., Hauser, T. U., Fearon, P., Bullmore, E. T., Goodyer, I. M., et al. (2019). Childhood socio-economic disadvantage predicts reduced myelin growth across adolescence and young adulthood. *bioRxiv*.
Zuo, X.-N., He, Y., Betzel, R. F., Colcombe, S., Sporns, O., & Milham, M. P. (2017). Human connectomics across the life span. *Trends in Cognitive Sciences*, *21*(1), 32–45.

Further reading

Hackman, D. A., & Farah, M. J. (2009). Socioeconomic status and the developing brain. *Trends in Cognitive Sciences*, *13*(2), 65–73.

CHAPTER FIVE

Psychosocial stress and epigenetic aging

Helena Palma-Gudiel[a,b], Lourdes Fañanás[a,b], Steve Horvath[c,d], Anthony S. Zannas[e,f,g,h,i,*]

[a]Department of Evolutionary Biology, Ecology and Environmental Sciences, Faculty of Biology, University of Barcelona (UB), Barcelona, Spain
[b]Centro de Investigación Biomédica en Red en Salud Mental (CIBERSAM), Madrid, Spain
[c]Department of Human Genetics, David Geffen School of Medicine, University of California Los Angeles, Los Angeles, CA, United States
[d]Department of Biostatistics, Fielding School of Public Health, University of California Los Angeles, Los Angeles, CA, United States
[e]Department of Psychiatry, University of North Carolina, Chapel Hill, NC, United States
[f]Department of Genetics, University of North Carolina, Chapel Hill, NC, United States
[g]Department of Psychiatry and Behavioral Sciences, Duke University Medical Center, Durham, NC, United States
[h]Institute for Trauma Recovery, University of North Carolina School of Medicine, Chapel Hill, NC, United States
[i]Neuroscience Curriculum, University of North Carolina, Chapel Hill, NC, United States
*Corresponding author: e-mail address: anthony_zannas@med.unc.edu

Contents

Abstract

Aging is the single most important risk factor for diseases that are currently the leading causes of morbidity and mortality. However, there is considerable inter-individual variability in risk for aging-related disease, and studies suggest that biological age can be influenced by multiple factors, including exposure to psychosocial stress. Among markers of biological age that can be affected by stress, the present article focuses on the so-called measures of epigenetic aging: DNA methylation-based age predictors that are measured in a range of tissues, including the brain, and can predict lifespan and healthspan. We review evidence linking exposure to diverse types of psychosocial stress, including early-life stress, cumulative stressful experiences, and low socioeconomic status, with accelerated epigenetic aging as a putative mediator of the effects of psychosocial environment on health and disease. The chapter also discusses methodological

International Review of Neurobiology, Volume 150
ISSN 0074-7742
https://doi.org/10.1016/bs.irn.2019.10.020

differences that may contribute to discordant findings across studies to date and plausible mechanisms that may underlie the effects of stress on the aging epigenome. Future studies examining the effects of adversity on epigenetic and other indicators of biological weathering may provide important insights into the pathogenesis of aging-related disease states.

1. Chronological versus biological age

Aging is the single most important risk factor for several disease states that are currently the leading causes of morbidity and mortality (Niccoli & Partridge, 2012). However, it is clear that health trajectories greatly differ among individuals of the same age, and biological age is thought to reflect weathering of the organism due to environmental exposures and disease processes. Efforts to better capture biological aging have led to the development of a wide range of biological measures, including telomere length, transcriptomic profiles, immunoglobulin G glycosylation patterns, blood serum metabolites, structural neuroimaging, and epigenomic profiles (Cole et al., 2018; Jylhävä, Pedersen, & Hägg, 2017). Among these measures, this chapter will focus on epigenetic markers, which are the most robust predictors of age developed to date (Horvath & Raj, 2018).

Epigenetics refers to any heritable chemical modification that affects gene expression without altering the genetic sequence itself. Epigenetic modifications are involved in several physiological processes such as the establishment and maintenance of cell lineages, X chromosome inactivation, and imprinting (Hackett & Azim Surani, 2013; Lee & Bartolomei, 2013). Furthermore, epigenetic mechanisms have been described to be involved in several complex disease states known to be influenced by gene–environment interactions (GxE). One of the most widely studied epigenetic modifications is DNA methylation, which comprises the addition of a methyl group to a cytosine residue and in humans most commonly occurs in the context of cytosine-guanine dinucleotides (CpG). DNA methylation changes can influence gene expression (Siegfried & Simon, 2010), thereby affecting gene and cell function.

Studies over the last years examining the role of age on epigenetic patterns have further led to the development of highly accurate epigenetic predictors of age or so-called epigenetic clocks. Table 1 summarizes the main characteristics of the four most widely used epigenetic clocks. Among the first ones and most widely used are those developed by Horvath (2013) and Hannum et al. (2013) in samples encompassing wide age ranges

Table 1 Characteristics of DNA methylation-based predictors of age.

	Horvath's DNAm Age	**Hannum's DNAm Age**	**PhenoAge (Levine et al.)**	**GrimAge (Lu et al.)**
CpG sites	353	71	513	1030
Platform[a]	27 K (450 K)	450 K	27 K (450 K, EPIC)	450 K (EPIC)
Training dataset (n)	3931	482	456	2356
Test dataset (n)	3211	174	9164	6935
Correlation with chronological age in testing dataset	0.96	0.91	0.62–0.89	0.82
Accuracy[b]	3.6 years	4.9 years	NA	NA
Tissue	51 different tissues	Whole blood	Whole blood	Whole blood
Age range	0–101	19–101	21–100	46–78

[a]DNA methylation array used for estimation; when some of the samples included in the study were assayed in larger platforms, they have been included inside parentheses to point out that actual CpG sites included in the regression model are solely the ones common to all platforms assayed.
[b]Accuracy refers to median absolute error (MAE), i.e., the median absolute difference between predicted epigenetic age and observed or chronological age.

(up to 100 years old). Both these clocks are characterized by their high accuracy (0.91–0.96 correlation between estimated and observed age), but Horvath's clock predicts age more accurately than Hannum's in samples of younger individuals. Furthermore, Horvath's clock can be used in different tissues and cell types since it was trained in 51 different tissues, including blood, saliva and different brain areas, whereas Hannum's clock was developed using blood samples only.

Interestingly, the estimation of epigenetic age allows the calculation of the so-called epigenetic age acceleration, which reflects an increase in biological age in comparison to chronological age of the individual. This acceleration has been associated with several physiological processes, pathologies, and environmental factors, including puberty (Binder et al., 2018), menopause (Levine et al., 2016), Alzheimer's disease (Levine, Lu, Bennett, & Horvath, 2015), and lifestyle (Quach et al., 2017), among several others (Gassen, Chrousos, Binder, & Zannas, 2016; Horvath & Raj, 2018). Exploration of the performance of the epigenetic clock in several tissues from the same subjects revealed that the cerebellum ages slowly when compared to other brain regions suggesting that different brain regions may be characterized by different rates of biological aging (Horvath et al., 2015).

Nonetheless, an epigenetic clock that is highly accurate in its age estimation might not be optimal for capturing inter-individual differences in epigenetic age acceleration. Thus, more recently developed clocks incorporate other parameters besides chronological age to better predict unfavorable outcomes. PhenoAge (Levine et al., 2018) integrates in the regression model a combination of clinical biomarkers such as blood concentration of albumin, creatinine, glucose, and C-reactive protein, and blood cell count. GrimAge (Lu et al., 2019) follows a similar approach by combining methylation levels at CpG sites which independently predict chronological age but also nine additional biomarkers including sex, leptin, cystatin C, and smoking pack years. In contrast with Horvath's and Hannum's age predictors, PhenoAge and GrimAge aim to predict healthspan rather than lifespan.

2. Epigenetic embedding of psychosocial stress

Because acceleration of epigenetic age has been repeatedly associated with mortality (time-to-death) (Breitling et al., 2016; Chen, Marioni, et al., 2016; Marioni et al., 2015; Perna et al., 2016), it is of interest to elucidate which mechanisms might accelerate, or potentially decelerate, epigenetic aging. Studies on monozygotic twins (MZ) suggest that environmental factors can contribute to DNA methylation changes and cumulative epigenetic differences along the lifespan (Fraga et al., 2005). While several environmental factors, such as lifestyle parameters and communicable diseases, have been associated with epigenetic age acceleration (Kananen et al., 2015; Quach et al., 2017; Rosen et al., 2018), this chapter focuses on the relationship between psychosocial stress and epigenetic aging.

Psychosocial stress here refers to any situation or event that results in emotional discomfort and/or is experienced as threatening by the exposed subject. Childhood maltreatment, domestic violence, the death of a loved one, migration, and low socioeconomic position (SEP) are among the most studied stressors in the scientific literature. Experiencing any of those, particularly during the early stages of life (Marín, 2016), has been reliably associated with an increased vulnerability for complex disorders, including psychiatric phenotypes (Moya-Higueras et al., 2018; Teicher & Samson, 2013). Exposure to psychosocial stressors such as adverse childhood experiences has been further associated with premature mortality; for example, subjects exposed to six or more childhood adversities were found to die on average 18.5 years earlier than subjects not exposed to adversity (Brown et al., 2009). Epigenetic mechanisms have been suggested as plausible mediators of this association (Gassen et al., 2016; Zannas et al., 2019).

Unlike biological parameters, exposure to psychosocial stress is more subjective and its assessment depends on memory recall. A recent meta-analysis demonstrated overall low agreement between prospective and retrospective measures of childhood maltreatment (Baldwin, Reuben, Newbury, & Danese, 2019). In the same study, information collected through interviews rather than questionnaires appeared to be more reliable. Likewise, different types of stress are likely to differentially affect biological and clinical outcomes depending on the frequency, type, severity, and timing of the particular event (Miller, Chen, & Zhou, 2007; Schalinski et al., 2016). Most studies assessing the effects of childhood trauma are conducted in adult populations by means of retrospective instruments; the Childhood Trauma Questionnaire (CTQ) is one of the most used throughout the literature, though it does not distinguish precise timing of the abuse (Bernstein et al., 1994).

Being born into an environment of low socioeconomic status (SES) is a risk factor for lifelong exposure to psychosocial stress (Lê-Scherban et al., 2018), since it has been associated with poorer parenting (Odgers et al., 2012) and increased exposure to stressful life events and violence (Lantz, House, Mero, & Williams, 2005). Furthermore, the impact of low SES has been suggested to moderate the effects of exposure to childhood maltreatment, but the underlying mechanisms remain to be elucidated (Goldberg et al., 2013). In this regard, lower SES is reliably associated with worse general health, poorer cognitive performance, and higher proneness for psychopathology (Aartsen et al., 2019; de Mestral & Stringhini, 2017; Lund et al., 2010).

Epigenetic mechanisms have been proposed to mediate the pernicious effects of the aforementioned environmental and sociodemographic risk factors. Genome-wide DNA methylation studies have revealed epigenetic patterns associated with socioeconomic status (Bush et al., 2018), institutionalization (Naumova et al., 2019), and altered stress reactivity after exposure to childhood maltreatment (Houtepen et al., 2016).

In this context, the aim of this chapter is to review evidence on whether psychosocial stress accelerates ticking of the epigenetic clock, potentially mediating known effects of stress exposure on health and disease outcomes. A comprehensive literature search yielded 15 scientific papers examining associations of DNA methylation-predicted age with exposure to psychosocial stress; their main findings are summarized in Table 2 and discussed in detail in the following sections.

Table 2 Studies examining epigenetic age acceleration after exposure to stress or trauma.

Study	Sample size	Mean age (SD; range)	Female sex (%)	Type of stress or trauma (instrument)	Clock used	Chronological/biological age correlation (r)	Tissue assessed	Main findings
Boks et al. (2015)	96	27.04 (9.15)	0	Traumatic stress (ETI)	Horvath	0.99	Whole blood	Higher exposure to (recent) combat trauma during deployment—but not childhood trauma—was associated with accelerated epigenetic aging
Zannas et al. (2015)	393	41.33 (12.85) 39.5 (14.14)	70.7 35.5	Childhood trauma (CTQ) Lifetime stress (SEQ)	Horvath	0.90 0.94	Whole blood	Cumulative life stress—but not current or childhood stress—was associated with accelerated epigenetic aging
Brody, Yu, Chen, Beach, and Miller (2016)	399	20	55	Parental depression (CES-D) and harsh parenting (4-item personalized interview)	Horvath	NA	PBMC	Parental depression (at age 11) predicted higher epigenetic age acceleration at age 20
Brody, Miller, Yu, Beach, and Chen (2016)	616	20–22	60.2	Racial discrimination during adolescence (schedule of racist events)	Hannum	NA	PBMC	In the absence of family support, exposure to racial discrimination is associated with accelerated epigenetic aging

Chen, Miller, et al. (2016)	330	19.3 (0.67)	53	SES (income, adequacy of income, unemployment, TANF receipt, unmet material needs, inability to make ends meet)	Horvath Hannum	NA	PBMC	Increased exposure to economic hardship during adolescence was associated with accelerated epigenetic aging in subjects exposed to the Great Recession
Simons et al. (2016)	100	48.5 (9.2)	100	SES (income and financial pressure) Childhood trauma (personalized 5-item interview) Lifestyle (smoking, alcohol intake, exercise, diet, BMI)	Hannum	0.82	PBMC	Lower income and higher financial pressure were associated with accelerated epigenetic aging. Neither childhood trauma nor lifestyle were associated with epigenetic age.
Jovanovic et al. (2017)	93	9.73 (1.67; 6–13)	54.5	Exposure to violence (VEX-R)	Horvath	0.46	Saliva	Exposure to violence was associated with accelerated epigenetic aging; this effect was driven by direct exposure rather than witnessing violence
Fiorito et al. (2017)	5111	57.3	48.0	SES (educational attainment, occupational position, income)	Horvath Hannum	0.73 to 0.80 depending on clock	Peripheral blood (whole blood, dried blood spots and PBMCs)	Lower SES was associated with accelerated epigenetic age

Continued

Table 2 Studies examining epigenetic age acceleration after exposure to stress or trauma.—cont'd

Study	Sample size	Mean age (SD; range)	Female sex (%)	Type of stress or trauma (instrument)	Clock used	Chronological/biological age correlation (r)	Tissue assessed	Main findings
Austin et al. (2018)	335	[15–55]	54.8	Early-life SES (parental occupational status) Current SES (maximum household occupational status)	Horvath	0.93	Monocytes	Low early-life SES (but not current) was associated with accelerated epigenetic age
Han et al. (2018)	811	41.5 (12.26)	66.7	Childhood trauma (NEMESIS-based childhood trauma interview)	Tailored for MBD-seq data	NA	Peripheral blood	Childhood trauma was associated with accelerated epigenetic aging
Hughes et al. (2018)	1099	58.4 (14.9; 28–98)	57.6	SES (income, employment, education, childhood SES as parents' social classification)	Horvath Hannum	0.90 0.94	Whole blood	Social disadvantage during childhood was associated with accelerated epigenetic age as measured by both Horvath and Hannum clocks. No associations were found for current SES
Lawn et al. (2018)	989 (twice) +773	28.65 (5.54) 47.44 (4.42) 53.44 (0.16)	100	SES (father's and highest current occupational social class) Psychosocial adversity during childhood (personalized ALSPAC/NSHD interview)	Horvath	0.45 0.50 0.12	Whole blood Buccal cells	Sexual abuse was associated with accelerated epigenetic aging in one of the cohorts at both time points. No associations were found for any other psychosocial or socioeconomic parameter

Sumner, Colich, Uddin, Armstrong, and McLaughlin (2019)	247	12.7 (2.6; 8–16)	47.8	Early-life adversity (categorized as threat[a] and deprivation[b])	Horvath	0.62	Saliva	Exposure to threat—but not to deprivation—was associated with accelerated epigenetic aging
Wolf et al. (2018)	2185	40.2 (18–89)	38.2	Childhood trauma (CTQ/ETI) and lifetime trauma (TLEQ, LEC, TEI, self-report)	Horvath Hannum	0.87 0.87	Whole blood	Childhood trauma but not lifetime trauma was associated with accelerated epigenetic aging as measured by Hannum's but not Horvath's clock
McCrory et al. (2019)	490	62.2 (8.3; 50–87)	50.2	SES (social class trajectory, education and income)	Horvath Hannum Levine	0.75 0.75 0.84	Whole blood	Epigenetic age acceleration was not associated with socioeconomic position

[a]Threat was conceptualized as an exposure composite including childhood abuse (either sexual, physical or emotional) and other threatening situations such as witnessing domestic violence.
[b]Deprivation was categorized as an exposure composite including childhood neglect (both emotional and physical), food insecurity and cognitive deprivation.
Abbreviations: ETI, Early Trauma Inventory; CES-D, Center for Epidemiological Studies Depression scale; CTQ, Childhood Trauma Questionnaire; LEC, Life Events Checklist; MBD-seq, Methyl-CpG Binding Domain Protein-Enriched Genome Sequencing; NA, not available; NEMESIS, Netherlands Mental Health Survey and Incidence Study; PBMC, peripheral blood mononuclear cells; SEQ, Stressful Events Questionnaire; SES, socioeconomic status; TANF, Temporary Assistance of Needy Families; TEI, Traumatic Events Inventory; TLEQ, Traumatic Life Events Questionnaire; VEX-R, Violence Exposure Scale for Children-Revised.

3. Exposure to life adversity accelerates epigenetic aging

Ten out of 15 studies examined different types of childhood trauma and maltreatment. Of those, seven reported significant associations with accelerated epigenetic aging. Interestingly, the only two studies performed in children and adolescents (from 6 to 16 years) yielded positive results suggesting that proximal experiences correlate with epigenetic changes (Jovanovic et al., 2017; Sumner et al., 2019). Both studies highlighted that the nature of the stressor influenced its impact on biological age, since (i) direct exposure rather than witnessing violence and (ii) exposure to threatening rather than depriving environments, were the specific factors triggering age acceleration, respectively. Nevertheless, studies assessing epigenetic age in children and adolescents should be examined with caution as they include samples with narrow age range, whereby measures of epigenetic age only modestly correlate with chronological age (Jovanovic et al., 2017; Simpkin et al., 2017; Sumner et al., 2019). In an effort to address these limitations, two new DNA methylation-based clocks have been developed in independent samples of children and adolescents to more accurately estimate epigenetic age in these younger populations (Li et al., 2018; McEwen et al., 2019).

With regard to adult samples in which childhood adversity was assessed retrospectively, a study found that only sexual abuse was associated with age acceleration as opposed to other forms of early psychosocial adversity such as parental death, adoption, or neglect (Lawn et al., 2018), further highlighting the differential role of childhood trauma depending on stressor type. Although Lawn et al. did not find any association with parental mental illness, another study reported accelerated epigenetic age after exposure to parental depression at age 11 (Brody, Yu, et al., 2016). Discrepancies between studies could arise from sociodemographic differences between assessed samples, since Lawn's study was conducted in middle-aged Caucasian women while Brody et al. assessed African American youths. In the African American population, supportive parenting was further associated with buffering of the pernicious effects of exposure to racial discrimination during adolescence (Brody, Miller, et al., 2016).

The effect of childhood maltreatment on epigenetic age acceleration has been supported by a recent meta-analysis, which found that CTQ-measured maltreatment was associated with accelerated epigenetic aging assessed with the Hannum but not the Horvath clock (Wolf et al., 2018). Nevertheless, neither childhood trauma measured with Traumatic Life Events Questionnaire (TLEQ) nor lifetime trauma were associated with epigenetic age acceleration, and generalizability of these findings may be limited because the study was conducted in patients with post-traumatic stress disorder (PTSD). In line with these findings in PTSD, Han et al. reported a positive association between childhood trauma and accelerated epigenetic age in adult depressed subjects (Han et al., 2018). In contrast, three studies did not find significant associations between childhood maltreatment and epigenetic age; specifically, one of them was conducted in recently deployed soldiers (Boks et al., 2015), whereas the other two assessed the presence of childhood trauma in middle-aged African American participants of low socioeconomic status (Simons et al., 2016; Zannas et al., 2015).

These discordant findings could be in part explained by differences in the sociodemographic, clinical, and stressor characteristics across study cohorts. Exposure to acute stressors could induce transient epigenetic marks that are subject to change in the long-term and may thus not be detected if epigenetic aging is measured much later in life. The type and intensity of stress exposure may greatly influence its impact on subsequent epigenetic changes; for example, violent events are more likely to increase allostatic load. Likewise, adult subjects with current psychiatric diagnoses, such as major depression or PTSD, have likely been exposed to more adverse childhood environments (Scott, Smith, & Ellis, 2010). Additionally, when adjusting for threatening events, experiences of deprivation seemed to delay pubertal development, suggesting that childhood adversities such as neglect or physical illnesses could decelerate rather than accelerate epigenetic aging. Because different types of childhood trauma greatly overlap in the general population (Vachon, Krueger, Rogosch, & Cicchetti, 2015), it is challenging to disentangle the effects of co-occurring exposures.

Only three studies assessed later exposures to stressful life events, yielding discordant findings. While higher exposure to recent combat trauma (Boks et al., 2015) and higher cumulative lifetime stress (Zannas et al., 2015) were associated with accelerated epigenetic age, Wolf et al. (2018) reported no association between lifetime trauma and epigenetic age. Further longitudinal studies may help elucidate how life events occurring at different life stages differentially affect epigenetic aging and aging-related processes along the human lifespan.

4. Socioeconomic status and epigenetic aging

Seven studies have explored whether lower socioeconomic status (SES) was associated with epigenetic age acceleration. Across these studies, current SES was defined based on (i) income, (ii) occupational position, and (iii) educational attainment. Early SES was defined based on parental occupational status during childhood. Studies assessing SES at both time points also considered SES trajectories as a function of whether SES had increased, decreased, or remained stable from childhood to adulthood.

While some of these studies found lower SES to be associated with accelerated epigenetic aging (Chen, Miller, et al., 2016; Fiorito et al., 2017; Simons et al., 2016), others did not find significant associations (Austin et al., 2018; Hughes et al., 2018; Lawn et al., 2018; McCrory et al., 2019). These discrepancies may result from differences in sample demographics and definition of SES across studies. Notably, Chen, Miller, et al. (2016) examined the youngest sample consisting of adolescents exposed to the Great Recession during their transition to adulthood. Simons et al. (2016) assessed African American women of very low-income settings, and the positive finding observed in this sample could in part be explained by reproductive factors, given that the number of pregnancies has been positively associated with maternal epigenetic aging (Ryan et al., 2018). Although Fiorito et al. (2017) assessed different indicators of SES, their main analysis focused on educational attainment using a standardized score adjusted for gender, age, ethnicity, and recruitment center rather than a direct measure such as years of education. Indeed, higher educational attainment has been repeatedly associated with lower epigenetic aging (Levine et al., 2018; Lu et al., 2019; Quach et al., 2017). In contrast, studies reporting lack of association between SES and epigenetic aging are characterized by the use of a dichotomous occupational status as a proxy for SES (Austin et al., 2018; Hughes et al., 2018; Lawn et al., 2018; McCrory et al., 2019). Lastly, although McCrory et al. (2019) found no epigenetic age acceleration with neither childhood nor adulthood low SES, they found an association with higher allostatic load as measured with a multi-modal marker derived from physiological measurements.

Parental income and occupational position during childhood were used as proxies of early SES in some studies (Austin et al., 2018; Hughes et al., 2018; Lawn et al., 2018). Two of these studies point to the preeminent role of early SES as a driver of accelerated epigenetic age. These results are in line with prior evidence suggesting that childhood is a critical developmental window for the programming of risk for complex disorders (Pervanidou & Chrousos, 2018).

An alternative approach is to study SES trajectories from childhood to adulthood. Stably high economic hardship—stably low SES—and upward mobility from low to high SES were both associated with epigenetic age acceleration when compared with stably high SES and downward mobility (Austin et al., 2018; Chen, Miller, et al., 2016). Such findings highlight that economic resources available during childhood may influence the ticking rate of the epigenetic clock. Notwithstanding, lack of difference in epigenetic aging across SES trajectories has also been reported (McCrory et al., 2019). The fact that upward mobility did not positively influence epigenetic aging is in line with exploratory studies suggesting that resilience after trauma is associated with a lasting epigenetic footprint. Specifically, higher self-control in low-SES youth predicted lower rates of aggressive behavior and substance abuse but was associated with accelerated epigenetic age (Miller, Yu, Chen, & Brody, 2015). Likewise, in a sample of war veterans with PTSD, higher resilience scores—better coping strategies—were also associated with accelerated epigenetic age (Mehta et al., 2018). Collectively, these findings suggest a biological state of accelerated aging and higher allostatic load in subjects that successfully overcome adversities.

Demographic characteristics other than SES should also be considered as potential confounders. A recent study explored how gender and ethnicity can influence the ticking rate of the epigenetic clock revealing men age faster than women and Caucasian, Hispanic, Amerindian and African American subjects show distinctive epigenetic aging patterns (Horvath et al., 2016).

5. Developmental programming of epigenetic age at birth

Interestingly, two additional DNA methylation-based epigenetic clocks were developed for the estimation of gestational age (GA) at birth (Bohlin et al., 2016; Knight et al., 2016; see Table 3 for further details). These clocks allow researchers to explore which prenatal risk factors might be influencing developmental maturity of newborns.

Intuitively, it could be hypothesized that higher prenatal stress would be associated with accelerated epigenetic GA; however, findings reported so far suggest an opposite direction of effects. Specifically, gestational diabetes in a

Table 3 Characteristics of DNA methylation-based predictors of gestational age.

	Knight's clock for gestational age	Bohlin's clock for gestational age
CpG sites	148	96/58
Platform[a]	27K (450K)	450K
Training dataset (n)	207	1068
Test dataset (n)	1135	685
Correlation with chronological age	0.91	0.81/0.7
Accuracy[b]	1.24 weeks	12.5/14.9 days
Tissue	Cord blood	Cord blood
Age range	24–42 gestational weeks	$\overline{x} = 40$ gestational weeks

[a]DNA methylation array used for estimation; when some of the samples included in the study were assayed in larger platforms, they have been included inside parentheses to point out that actual CpG sites included in the regression model are solely the ones common to all platforms assayed.
[b]Accuracy refers to median absolute error (MAE), i.e., the median absolute difference between predicted epigenetic age and observed or chronological age.

previous pregnancy, Sjögren's syndrome, maternal history of depression before pregnancy, antenatal depressive symptoms, and lower cerebroplacental ratio during pregnancy (a proxy for prenatal hypoxia and/or placental resistance) were all associated with epigenetic GA deceleration (Girchenko et al., 2017; Palma-Gudiel et al., 2019; Suarez et al., 2018). Thus, exposure to stress during prenatal stages of life is associated with decreased biological age, which could be conceptualized as a marker of developmental immaturity at birth. Knight et al. (2018) further tested the clinical relevance of epigenetic GA deceleration at birth in terms of perinatal health revealing an increased need of neonatal interventions and heightened risk of bronchopulmonary dysplasia. Conversely, excessive maternal weight and obesity before pregnancy were associated with accelerated epigenetic GA at birth (Khouja et al., 2018), suggesting that these conditions hasten fetal development. Girchenko et al. (2017) also found maternal pre-eclampsia, maternal age over 40 years at delivery, and fetal demise in a previous pregnancy to be associated with accelerated epigenetic GA.

Taken together, these results suggest a dual nature for the effects of stress on aging-related processes, whereby depriving environments may give rise to delayed development—and thus decelerated epigenetic age—whereas threatening events may be associated with heightened allostatic load and accelerated epigenetic aging. Notably, the rate of change of epigenetic aging is faster during development (Horvath & Raj, 2018), and while age acceleration might be beneficial during the early stages of life—promoting survival and growth—it may be also associated with later premature mortality. Large epidemiological and longitudinal studies should be conducted to explore how the rate of biological aging may impact developmental programming. Furthermore, alternative measures of biological aging, such as neuroimaging-derived brain age, have revealed the positive association between prenatal undernutrition and premature brain aging (Franke, Gaser, Roseboom, Schwab, & de Rooij, 2018). These results highlight the interest of exploring simultaneously different markers of biological age, as different organs are known to have independent age rates in the same individuals; thus, there could be tissue-specific effects of epigenetic age predictors, e.g., brain epigenetic aging better informs about neurodegenerative processes (Levine et al., 2015) while blood epigenetic age better reflects cardiovascular risk (Huang et al., 2019).

6. Mechanistic insights of stress-mediated accelerated aging

Throughout evolution, living beings have developed diverse strategies to cope with stress. In vertebrates and humans, the hypothalamic-pituitary-adrenal (HPA) axis culminates with cortisol secretion in the bloodstream upon exposure to stressful stimuli. As a lipophilic molecule, cortisol diffuses through the cell membrane and reaches the cytoplasm where it can bind to the glucocorticoid receptor (GR). Once bound to cortisol, the GR translocates to the cell nucleus to regulate gene expression. The key role played by the GR in the stress response has prompted abundant research on how stress exposure could lead to altered DNA methylation of the *NR3C1* gene encoding this receptor (Palma-Gudiel, Córdova-Palomera, Leza, & Fañanás, 2015). Specifically, the GR exerts its actions through binding to so-called glucocorticoid responsive elements (GREs) (Palma-Gudiel et al., 2018). Notably, a large proportion of CpG sites comprising the Horvath's epigenetic clock co-localize with GREs and are susceptible to glucocorticoid exposure, suggesting that circulating glucocorticoids could mediate the effects of psychosocial stress on epigenetic aging (Zannas et al., 2015).

Beyond such composite epigenomic effects, stress may also impart epigenetic changes at genomic loci relevant for aging-related phenomena.

Such a locus is *FKBP5*, the gene encoding a stress-responsive molecule implicated in regulation of HPA axis function and stress responses. Previously, *FKBP5* DNA methylation was shown to partially mediate the known association between exposure to childhood trauma and later development of PTSD (Klengel et al., 2013). More recently, epigenetic upregulation of *FKBP5* was found to result from the synergistic effects of aging and stress and to contribute to increased inflammation and cardiovascular risk (Zannas et al., 2019). These findings suggest that epigenetic regulation of FKBP5, and likely other stress-responsive molecules, may in part mediate the impact of psychosocial stress on aging-related disease states.

Elucidating the mechanisms through which stress exposure contributes to accelerated epigenetic aging may enhance our ability to prevent the high mortality and morbidity rates associated with adverse environments. Predictions based on DNA methylation-based markers may also inform which interventions have the potential to buffer the pernicious effects of stress at the molecular level. Evidence to date suggests that early-life stress can be deleterious, but longitudinal studies are needed to clarify the dynamics and putative stability or reversibility of the effects on epigenetic aging. The advent of novel measures of epigenetic aging aimed at predicting not only lifespan but also healthspan opens new avenues for research in the field. Future studies examining the effects of adversity on these indicators of biological weathering may provide novel insights into aging-related phenomena. Postmortem brain analyses are also needed to understand how different kinds of stress can accelerate epigenetic aging not only in peripheral tissues such as blood or saliva but also in different brain areas as it seems to be associated with several neurodegenerative conditions.

References

Aartsen, M. J., Cheval, B., Sieber, S., Van der Linden, B. W., Gabriel, R., Courvoisier, D. S., et al. (2019). Advantaged socioeconomic conditions in childhood are associated with higher cognitive functioning but stronger cognitive decline in older age. *Proceedings of the National Academy of Sciences*, *116*(12), 5478 LP–5486. https://doi.org/10.1073/pnas.1807679116.

Austin, M. K., Chen, E., Ross, K. M., McEwen, L. M., Maclsaac, J. L., Kobor, M. S., et al. (2018). Early-life socioeconomic disadvantage, not current, predicts accelerated epigenetic aging of monocytes. *Psychoneuroendocrinology*, *97*(January), 131–134. https://doi.org/10.1016/j.psyneuen.2018.07.007.

Baldwin, J. R., Reuben, A., Newbury, J. B., & Danese, A. (2019). Agreement between prospective and retrospective measures of childhood maltreatment: A systematic review and meta-analysis. *JAMA Psychiatry*, *76*(6), 584–593. https://doi.org/10.1001/jamapsychiatry.2019.0097.

Bernstein, D. P., Fink, L., Handelsman, L., Foote, J., Lovejoy, M., Wenzel, K., et al. (1994). Initial reliability and validity of a new retrospective measure of child abuse and neglect. *American Journal of Psychiatry*, *151*(8), 1132–1136. https://doi.org/10.1176/ajp.151.8.1132.

Binder, A. M., Corvalan, C., Mericq, V., Pereira, A., Santos, J. L., Horvath, S., et al. (2018). Faster ticking rate of the epigenetic clock is associated with faster pubertal development in girls. *Epigenetics*, *13*(1), 85–94. https://doi.org/10.1080/15592294.2017.1414127.

Bohlin, J., Håberg, S. E., Magnus, P., Reese, S. E., Gjessing, H. K., Magnus, M. C., et al. (2016). Prediction of gestational age based on genome-wide differentially methylated regions. *Genome Biology*, *17*(1), 207. https://doi.org/10.1186/s13059-016-1063-4.

Boks, M. P., van Mierlo, H. C., Rutten, B. P. F., Radstake, T. R. D. J., De Witte, L., Geuze, E., et al. (2015). Longitudinal changes of telomere length and epigenetic age related to traumatic stress and post-traumatic stress disorder. *Psychoneuroendocrinology*, *51*, 506–512. https://doi.org/10.1016/j.psyneuen.2014.07.011.

Breitling, L. P., Saum, K.-U., Perna, L., Schöttker, B., Holleczek, B., & Brenner, H. (2016). Frailty is associated with the epigenetic clock but not with telomere length in a German cohort. *Clinical Epigenetics*, *8*(1), 21. https://doi.org/10.1186/s13148-016-0186-5.

Brody, G. H., Miller, G. E., Yu, T., Beach, S. R. H., & Chen, E. (2016). Supportive family environments ameliorate the link between racial discrimination and epigenetic aging: A replication across two longitudinal cohorts. *Psychological Science*, *27*(4), 530–541. https://doi.org/10.1177/0956797615626703.

Brody, G. H., Yu, T., Chen, E., Beach, S. R. H., & Miller, G. E. (2016). Family-centered prevention ameliorates the longitudinal association between risky family processes and epigenetic aging. *Journal of Child Psychology and Psychiatry, and Allied Disciplines*, *57*(5), 566–574. https://doi.org/10.1111/jcpp.12495.

Brown, D. W., Anda, R. F., Tiemeier, H., Felitti, V. J., Edwards, V. J., Croft, J. B., et al. (2009). Adverse childhood experiences and the risk of premature mortality. *American Journal of Preventive Medicine*, *37*(5), 389–396. https://doi.org/10.1016/j.amepre.2009.06.021.

Bush, N. R., Edgar, R. D., Park, M., MacIsaac, J. L., McEwen, L. M., Adler, N. E., et al. (2018). The biological embedding of early-life socioeconomic status and family adversity in children's genome-wide DNA methylation. *Epigenomics*, *10*(11), 1445–1461. https://doi.org/10.2217/epi-2018-0042.

Chen, B. H., Marioni, R. E., Colicino, E., Peters, M. J., Ward-Caviness, C. K., Tsai, P. C., et al. (2016). DNA methylation-based measures of biological age: Meta-analysis predicting time to death. *Aging*, *8*, 1844–1865. https://doi.org/10.18632/aging.101020.

Chen, E., Miller, G. E., Yu, T., & Brody, G. H. (2016). The Great Recession and health risks in African American youth. *Brain, Behavior, and Immunity*, *53*, 234–241. https://doi.org/10.1016/j.bbi.2015.12.015.

Cole, J. H., Ritchie, S. J., Bastin, M. E., Valdés Hernández, M. C., Muñoz Maniega, S., Royle, N., et al. (2018). Brain age predicts mortality. *Molecular Psychiatry*, *23*(5), 1385–1392. https://doi.org/10.1038/mp.2017.62.

de Mestral, C., & Stringhini, S. (2017). Socioeconomic status and cardiovascular disease: An update. *Current Cardiology Reports*, *19*(11), 115. https://doi.org/10.1007/s11886-017-0917-z.

Fiorito, G., Polidoro, S., Dugué, P. A., Kivimaki, M., Ponzi, E., Matullo, G., et al. (2017). Social adversity and epigenetic aging: A multi-cohort study on socioeconomic differences in peripheral blood DNA methylation. *Scientific Reports*, 7(1), 1–12. https://doi.org/10.1038/s41598-017-16391-5.

Fraga, M. F., Ballestar, E., Paz, M. F., Ropero, S., Setien, F., Ballestar, M. L., et al. (2005). Epigenetic differences arise during the lifetime of monozygotic twins. *Proceedings of the National Academy of Sciences of the United States of America*, *102*(30), 10604–10609. https://doi.org/10.1073/pnas.0500398102.

Franke, K., Gaser, C., Roseboom, T. J., Schwab, M., & de Rooij, S. R. (2018). Premature brain aging in humans exposed to maternal nutrient restriction during early gestation. *NeuroImage*, *173*, 460–471. https://doi.org/10.1016/j.neuroimage.2017.10.047.

Gassen, N. C., Chrousos, G. P., Binder, E. B., & Zannas, A. S. (2016). Life stress, glucocorticoid signaling, and the aging epigenome: Implications for aging-related diseases. *Neuroscience and Biobehavioral Reviews*, *74*(Pt. B), 356–365. https://doi.org/10.1016/j.neubiorev.2016.06.003.

Girchenko, P., Lahti, J., Czamara, D., Knight, A. K., Jones, M. J., Suarez, A., et al. (2017). Associations between maternal risk factors of adverse pregnancy and birth outcomes and the offspring epigenetic clock of gestational age at birth. *Clinical Epigenetics*, *9*(1), 1–14. https://doi.org/10.1186/s13148-017-0349-z.

Goldberg, X., Alemany, S., Fatjó-Vilas, M., González-Ortega, I., González-Pinto, A., Cuesta, M. J., et al. (2013). Twin-based study of the complex interplay between childhood maltreatment, socioeconomic status and adult memory. *European Archives of Psychiatry and Clinical Neuroscience*, *263*(5), 435–440. https://doi.org/10.1007/s00406-012-0382-z.

Hackett, J. A., & Azim Surani, M. (2013). DNA methylation dynamics during the mammalian life cycle. *Philosophical Transactions of the Royal Society, B: Biological Sciences*, *368*(1609), 1–8. https://doi.org/10.1098/rstb.2011.0328.

Han, L. K. M., Aghajani, M., Clark, S. L., Chan, R. F., Hattab, M. W., Shabalin, A. A., et al. (2018). Epigenetic aging in major depressive disorder. *American Journal of Psychiatry*, *175*(8), 774–782. https://doi.org/10.1176/appi.ajp.2018.17060595.

Hannum, G., Guinney, J., Zhao, L., Zhang, L., Hughes, G., Sadda, S. V., et al. (2013). Genome-wide methylation profiles reveal quantitative views of human aging rates. *Molecular Cell*, *49*(2), 359–367. https://doi.org/10.1016/j.molcel.2012.10.016.

Horvath, S. (2013). DNA methylation age of human tissues and cell types. *Genome Biology*, *14*(10), R115. https://doi.org/10.1186/gb-2013-14-10-r115.

Horvath, S., Gurven, M., Levine, M. E., Trumble, B. C., Kaplan, H., Allayee, H., et al. (2016). An epigenetic clock analysis of race/ethnicity, sex, and coronary heart disease. *Genome Biology*, *17*(1), 171. https://doi.org/10.1186/s13059-016-1030-0.

Horvath, S., Mah, V., Lu, A. T., Woo, J. S., Choi, O. W., Jasinska, A. J., et al. (2015). The cerebellum ages slowly according to the epigenetic clock. *Aging*, 7(5), 294–305.

Horvath, S., & Raj, K. (2018). DNA methylation-based biomarkers and the epigenetic clock theory of ageing. *Nature Reviews Genetics*, *19*(6), 371–384. https://doi.org/10.1038/s41576-018-0004-3.

Houtepen, L. C., Vinkers, C. H., Carrillo-Roa, T., Hiemstra, M., van Lier, P. A., Meeus, W., et al. (2016). Genome-wide DNA methylation levels and altered cortisol stress reactivity following childhood trauma in humans. *Nature Communications*, 7, 10967. Retrieved from https://doi.org/10.1038/ncomms10967.

Huang, R.-C., Lillycrop, K. A., Beilin, L. J., Godfrey, K. M., Anderson, D., Mori, T. A., et al. (2019). Epigenetic age acceleration in adolescence associates with BMI, inflammation, and risk score for middle age cardiovascular disease. *The Journal of Clinical Endocrinology & Metabolism*, *104*(7), 3012–3024. https://doi.org/10.1210/jc.2018-02076.

Hughes, A., Smart, M., Gorrie-Stone, T., Hannon, E., Mill, J., Bao, Y., et al. (2018). Socioeconomic position and DNA methylation age acceleration across the life course. *American Journal of Epidemiology*, *187*(11), 2346–2354. https://doi.org/10.1093/aje/kwy155.

Jovanovic, T., Vance, L. A., Cross, D., Knight, A. K., Kilaru, V., Michopoulos, V., et al. (2017). Exposure to violence accelerates epigenetic aging in children. *Scientific Reports*, 7(1), 1–7. https://doi.org/10.1038/s41598-017-09235-9.

Jylhävä, J., Pedersen, N. L., & Hägg, S. (2017). Biological age predictors. *eBioMedicine*, *21*, 29–36. https://doi.org/10.1016/j.ebiom.2017.03.046.

Kananen, L., Nevalainen, T., Jylhävä, J., Marttila, S., Hervonen, A., Jylhä, M., et al. (2015). Cytomegalovirus infection accelerates epigenetic aging. *Experimental Gerontology*, *72*, 227–229. https://doi.org/10.1016/j.exger.2015.10.008.

Khouja, J. N., Simpkin, A. J., O'Keeffe, L. M., Wade, K. H., Houtepen, L. C., Relton, C. L., et al. (2018). Epigenetic gestational age acceleration: A prospective cohort study investigating associations with familial, sociodemographic and birth characteristics. *Clinical Epigenetics*, *10*(1), 86. https://doi.org/10.1186/s13148-018-0520-1.

Klengel, T., Mehta, D., Anacker, C., Rex-Haffner, M., Pruessner, J. C., Pariante, C. M., et al. (2013). Allele-specific FKBP5 DNA demethylation mediates gene-childhood trauma interactions. *Nature Neuroscience*, *16*(1), 33–41. https://doi.org/10.1038/nn.3275.

Knight, A. K., Craig, J. M., Theda, C., Bækvad-Hansen, M., Bybjerg-Grauholm, J., Hansen, C. S., et al. (2016). An epigenetic clock for gestational age at birth based on blood methylation data. *Genome Biology*, *17*(1), 1–11. https://doi.org/10.1186/s13059-016-1068-z.

Knight, A. K., Smith, A. K., Conneely, K. N., Dalach, P., Loke, Y. J., Cheong, J. L., et al. (2018). Relationship between epigenetic maturity and respiratory morbidity in preterm infants. *The Journal of Pediatrics*, *198*, 168–173.e2. https://doi.org/10.1016/j.jpeds.2018.02.074.

Lantz, P. M., House, J. S., Mero, R. P., & Williams, D. R. (2005). Stress, life events, and socioeconomic disparities in health: Results from the Americans' changing lives study. *Journal of Health and Social Behavior*, *46*(3), 274–288. https://doi.org/10.1177/002214650504600305.

Lawn, R. B., Anderson, E. L., Suderman, M., Simpkin, A. J., Gaunt, T. R., Teschendorff, A. E., et al. (2018). Psychosocial adversity and socioeconomic position during childhood and epigenetic age: Analysis of two prospective cohort studies. *Human Molecular Genetics*, *27*(7), 1301–1308. https://doi.org/10.1093/hmg/ddy036.

Lee, J. T., & Bartolomei, M. S. (2013). X-inactivation, imprinting, and long noncoding RNAs in health and disease. *Cell*, *152*(6), 1308–1323. https://doi.org/10.1016/j.cell.2013.02.016.

Lê-Scherban, F., Brenner, A. B., Hicken, M. T., Needham, B. L., Seeman, T., Sloan, R. P., et al. (2018). Child and adult socioeconomic status and the cortisol response to acute stress: Evidence from the multi-ethnic study of atherosclerosis. *Psychosomatic Medicine*, *80*(2), 184–192. Retrieved from https://journals.lww.com/psychosomaticmedicine/Fulltext/2018/02000/Child_and_Adult_Socioeconomic_Status_and_the.8.aspx.

Levine, M. E., Lu, A. T., Bennett, D. A., & Horvath, S. (2015). Epigenetic age of the prefrontal cortex is associated with neuritic plaques, amyloid load, and Alzheimer's disease related cognitive functioning. *Aging*, 7(12), 1198–1211. https://doi.org/10.18632/aging.100864.

Levine, M. E., Lu, A. T., Chen, B. H., Hernandez, D. G., Singleton, A. B., Ferrucci, L., et al. (2016). Menopause accelerates biological aging. *Proceedings of the National Academy of Sciences of the United States of America*, *113*(33), 9327–9332. https://doi.org/10.1073/pnas.1604558113.

Levine, M. E., Lu, A. T., Quach, A., Chen, B. H., Assimes, T. L., Bandinelli, S., et al. (2018). An epigenetic biomarker of aging for lifespan and healthspan. *Aging*, *10*(4), 573–591. https://doi.org/10.18632/aging.101414.

Li, C., Gao, W., Gao, Y., Yu, C., Lv, J., Lv, R., et al. (2018). Age prediction of children and adolescents aged 6-17 years: An epigenome-wide analysis of DNA methylation. *Aging*, *10*(5), 1015–1026. https://doi.org/10.18632/aging.101445.

Lu, A. T., Quach, A., Wilson, J. G., Reiner, A. P., Aviv, A., Raj, K., et al. (2019). DNA methylation GrimAge strongly predicts lifespan and healthspan. *Aging*, *11*(2), 303–327. https://doi.org/10.18632/aging.101684.

Lund, C., Breen, A., Flisher, A. J., Kakuma, R., Corrigall, J., Joska, J. A., et al. (2010). Poverty and common mental disorders in low and middle income countries: A systematic review. *Social Science & Medicine*, *71*(3), 517–528. https://doi.org/10.1016/j.socscimed.2010.04.027.

Marín, O. (2016). Developmental timing and critical windows for the treatment of psychiatric disorders. *Nature Medicine*, *22*(11), 1229–1238. https://doi.org/10.1038/nm.4225.

Marioni, R. E., Shah, S., McRae, A. F., Chen, B. H., Colicino, E., Harris, S. E., et al. (2015). DNA methylation age of blood predicts all-cause mortality in later life. *Genome Biology*, *16*(1), 25. https://doi.org/10.1186/s13059-015-0584-6.

McCrory, C., Fiorito, G., Ni Cheallaigh, C., Polidoro, S., Karisola, P., Alenius, H., et al. (2019). How does socio-economic position (SEP) get biologically embedded? A comparison of allostatic load and the epigenetic clock(s). *Psychoneuroendocrinology*, *104*, 64–73. https://doi.org/10.1016/j.psyneuen.2019.02.018 [October 2018].

McEwen, L. M., O'Donnell, K. J., McGill, M. G., Edgar, R. D., Jones, M. J., MacIsaac, J. L., et al. (2019). The PedBE clock accurately estimates DNA methylation age in pediatric buccal cells. *Proceedings of the National Academy of Sciences of the United States of America*. https://doi.org/10.1073/pnas.1820843116.

Mehta, D., Bruenig, D., Lawford, B., Harvey, W., Carrillo-Roa, T., Morris, C. P., et al. (2018). Accelerated DNA methylation aging and increased resilience in veterans: The biological cost for soldiering on. *Neurobiology of Stress*, *8*, 112–119. https://doi.org/10.1016/j.ynstr.2018.04.001 [November 2017].

Miller, G. E., Chen, E., & Zhou, E. S. (2007). If it goes up, must it come down? Chronic stress and the hypothalamic-pituitary-adrenocortical axis in humans. *Psychological Bulletin*, *133*(1), 25–45. https://doi.org/10.1037/0033-2909.133.1.25.

Miller, G. E., Yu, T., Chen, E., & Brody, G. H. (2015). Self-control forecasts better psychosocial outcomes but faster epigenetic aging in low-SES youth. *Proceedings of the National Academy of Sciences of the United States of America*, *112*(33), 10325–10330. https://doi.org/10.1073/pnas.1505063112.

Moya-Higueras, J., Cuevas, A., Marques-Feixa, L., Mezquita, L., Mayoral, M., Fañanás, L., et al. (2018). Recent stressful life events (SLE) and adolescent mental health: Initial validation of the LEIA, a new checklist for SLE assessment according to their severity, interpersonal, and dependent nature. *Assessment*. 1073191118817648. https://doi.org/10.1177/1073191118817648.

Naumova, O. Y., Rychkov, S. Y., Kornilov, S. A., Odintsova, V. V., Anikina, V. O., Solodunova, M. Y., et al. (2019). Effects of early social deprivation on epigenetic statuses and adaptive behavior of young children: A study based on a cohort of institutionalized infants and toddlers. *PLoS One*, *14*(3), e0214285. Retrieved from https://doi.org/10.1371/journal.pone.0214285.

Niccoli, T., & Partridge, L. (2012). Ageing as a risk factor for disease. *Current Biology*, *22*(17), R741–R752. https://doi.org/10.1016/j.cub.2012.07.024.

Odgers, C. L., Caspi, A., Russell, M. A., Sampson, R. J., Arseneault, L., & Moffitt, T. E. (2012). Supportive parenting mediates neighborhood socioeconomic disparities in children's antisocial behavior from ages 5 to 12. *Development and Psychopathology*, *24*, 705–721. https://doi.org/10.1017/s0954579412000326.

Palma-Gudiel, H., Córdova-Palomera, A., Leza, J. C., & Fañanás, L. (2015). Glucocorticoid receptor gene (NR3C1) methylation processes as mediators of early adversity in stress-related disorders causality: A critical review. *Neuroscience and Biobehavioral Reviews*, *55*, 520–535. https://doi.org/10.1016/j.neubiorev.2015.05.016.

Palma-Gudiel, H., Córdova-Palomera, A., Tornador, C., Falcón, C., Bargalló, N., Deco, G., et al. (2018). Increased methylation at an unexplored glucocorticoid responsive element within exon 1 D of NR3C1 gene is related to anxious-depressive disorders and decreased hippocampal connectivity. *European Neuropsychopharmacology*, *28*(5), 579–588. https://doi.org/10.1016/j.euroneuro.2018.03.015.
Palma-Gudiel, H., Eixarch, E., Crispi, F., Morán, S., Zannas, A. S., & Fañanás, L. (2019). Prenatal adverse environment is associated with epigenetic age deceleration at birth and hypomethylation at the hypoxia-responsive EP300 gene. *Clinical Epigenetics*, *11*(1), 73. https://doi.org/10.1186/s13148-019-0674-5.
Perna, L., Zhang, Y., Mons, U., Holleczek, B., Saum, K.-U., & Brenner, H. (2016). Epigenetic age acceleration predicts cancer, cardiovascular, and all-cause mortality in a German case cohort. *Clinical Epigenetics*, *8*(1), 64. https://doi.org/10.1186/s13148-016-0228-z.
Pervanidou, P., & Chrousos, G. P. (2018). Early-life stress: From neuroendocrine mechanisms to stress-related disorders. *Hormone Research in Pædiatrics*, *89*(5), 372–379. https://doi.org/10.1159/000488468.
Quach, A., Levine, M. E., Tanaka, T., Lu, A. T., Chen, B. H., Ferrucci, L., et al. (2017). Epigenetic clock analysis of diet, exercise, education, and lifestyle factors. *Aging*, *9*(2), 419–446. https://doi.org/10.18632/aging.101168.
Rosen, A. D., Robertson, K. D., Hlady, R. A., Muench, C., Lee, J., Philibert, R., et al. (2018). DNA methylation age is accelerated in alcohol dependence. *Translational Psychiatry*, *8*(1), 182. https://doi.org/10.1038/s41398-018-0233-4.
Ryan, C. P., Hayes, M. G., Lee, N. R., McDade, T. W., Jones, M. J., Kobor, M. S., et al. (2018). Reproduction predicts shorter telomeres and epigenetic age acceleration among young adult women. *Scientific Reports*, *8*(1), 1–9. https://doi.org/10.1038/s41598-018-29486-4.
Schalinski, I., Teicher, M. H., Nischk, D., Hinderer, E., Müller, O., & Rockstroh, B. (2016). Type and timing of adverse childhood experiences differentially affect severity of PTSD, dissociative and depressive symptoms in adult inpatients. *BMC Psychiatry*, *16*(1), 295. https://doi.org/10.1186/s12888-016-1004-5.
Scott, K. M., Smith, D. R., & Ellis, P. M. (2010). Prospectively ascertained child maltreatment and its association with DSM-IV mental disorders in young adults. *Archives of General Psychiatry*, *67*, 712–719. https://doi.org/10.1001/archgenpsychiatry.2010.71.
Siegfried, Z., & Simon, I. (2010). DNA methylation and gene expression. *Wiley Interdisciplinary Reviews: Systems Biology and Medicine*, *2*(3), 362–371. https://doi.org/10.1002/wsbm.64.
Simons, R. L., Lei, M. K., Beach, S. R. H., Philibert, R. A., Cutrona, C. E., Gibbons, F. X., et al. (2016). Economic hardship and biological weathering: The epigenetics of aging in a U.S. sample of black women. *Social Science and Medicine*, *150*, 192–200. https://doi.org/10.1016/j.socscimed.2015.12.001.
Simpkin, A. J., Howe, L. D., Tilling, K., Gaunt, T. R., Lyttleton, O., McArdle, W. L., et al. (2017). The epigenetic clock and physical development during childhood and adolescence: Longitudinal analysis from a UK birth cohort. *International Journal of Epidemiology*, *46*(2), 549–558. https://doi.org/10.1093/ije/dyw307.
Suarez, A., Lahti, J., Czamara, D., Lahti-Pulkkinen, M., Knight, A. K., Girchenko, P., et al. (2018). The epigenetic clock at birth: Associations with maternal antenatal depression and child psychiatric problems. *Journal of the American Academy of Child and Adolescent Psychiatry*, *57*(5), 321–328.e2. https://doi.org/10.1016/j.jaac.2018.02.011.
Sumner, J. A., Colich, N. L., Uddin, M., Armstrong, D., & McLaughlin, K. A. (2019). Early experiences of threat, but not deprivation, are associated with accelerated biological aging in children and adolescents. *Biological Psychiatry*, *85*(3), 268–278. https://doi.org/10.1016/j.biopsych.2018.09.008.

Teicher, M. H., & Samson, J. A. (2013). Childhood maltreatment and psychopathology: A case for ecophenotypic variants as clinically and neurobiologically distinct subtypes. *The American Journal of Psychiatry*, *170*(10), 1114–1133. https://doi.org/10.1176/appi.ajp.2013.12070957.

Vachon, D. D., Krueger, R. F., Rogosch, F. A., & Cicchetti, D. (2015). Assessment of the harmful psychiatric and behavioral effects of different forms of child maltreatment. *JAMA Psychiatry*, *72*(11), 1135–1142. https://doi.org/10.1001/jamapsychiatry.2015.1792.

Wolf, E. J., Maniates, H., Nugent, N., Maihofer, A. X., Armstrong, D., Ratanatharathorn, A., et al. (2018). Traumatic stress and accelerated DNA methylation age: A meta-analysis. *Psychoneuroendocrinology*, *92*, 123–134. https://doi.org/10.1016/j.psyneuen.2017.12.007 (November 2017).

Zannas, A. S., Arloth, J., Carrillo-Roa, T., Iurato, S., Röh, S., Ressler, K. J., et al. (2015). Lifetime stress accelerates epigenetic aging in an urban, African American cohort: Relevance of glucocorticoid signaling. *Genome Biology*, *16*(1), 1–12. https://doi.org/10.1186/s13059-015-0828-5.

Zannas, A. S., Jia, M., Hafner, K., Baumert, J., Wiechmann, T., Pape, J. C., et al. (2019). Epigenetic upregulation of FKBP5 by aging and stress contributes to NF-κB–driven inflammation and cardiovascular risk. *Proceedings of the National Academy of Sciences of the United States of America*, *116*(23), 11370–11379. https://doi.org/10.1073/pnas..1816847116.

CHAPTER SIX

Effects of stress on the structure and function of the medial prefrontal cortex: Insights from animal models

Cara L. Wellman[a,b,c,*], Justin L. Bollinger[d], Kelly M. Moench[e]

[a]Department of Psychological & Brain Sciences, Indiana University, Bloomington, IN, United States
[b]Center for the Integrative Study of Animal Behavior, Indiana University, Bloomington, IN, United States
[c]Program in Neuroscience, Indiana University, Bloomington, IN, United States
[d]Department of Pharmacology & Systems Physiology, University of Cincinnati College of Medicine, Cincinnati, OH, United States
[e]Department of Psychology, Indiana University-Purdue University, Indianapolis, IN, United States
*Corresponding author: e-mail address: wellmanc@indiana.edu

Contents

Abstract

Stress alters both cognitive and emotional function, and increases risk for a variety of psychological disorders, such as depression and posttraumatic stress disorder. The prefrontal cortex is critical for executive function and emotion regulation, is a target for stress hormones, and is implicated in many stress-influenced psychological disorders. Therefore, understanding how stress-induced changes in the structure and function of the prefrontal cortex are related to stress-induced changes in behavior may elucidate some of the mechanisms contributing to stress-sensitive disorders. This review focuses on data from rodent models to describe the effects of chronic stress on behaviors mediated by the medial prefrontal cortex, the effects of chronic stress on the morphology and physiology of the medial prefrontal cortex, mechanisms that may mediate these

International Review of Neurobiology, Volume 150
ISSN 0074-7742
https://doi.org/10.1016/bs.irn.2019.11.007

effects, and evidence for sex differences in the effects of stress on the prefrontal cortex. Understanding how stress influences prefrontal cortex and behaviors mediated by it, as well as sex differences in this effect, will elucidate potential avenues for novel interventions for stress-sensitive disorders characterized by deficits in executive function and emotion regulation.

1. Introduction

Stressful experiences, which broadly can be considered to be challenges to an organism's wellbeing or allostasis (McEwen, 2017), can vary in their intensity, duration, and chronicity. Further, cognitive processing of potentially stressful events can vary across individuals. Thus, it is not surprising that such challenges can have a range of effects, varying from beneficial to detrimental (Sapolsky, 2003). Nonetheless, excessive, chronic, or repeated stress can produce cognitive and emotional dysfunction. For instance, stressful life events can precipitate episodes of major depression (Risch et al., 2009), traumatic stressors can trigger posttraumatic stress disorder (Lechin, Van der Dijs, & Benaim, 1996; Turner & Lloyd, 2004), and risk for mood and anxiety disorders increases with repeated exposure to stress (Risch et al., 2009). Such chronic stressors have adverse effects on many behaviors in the absence of overt psychopathology. For instance, disrupted frontoparietal connectivity and associated deficits in attentional control have been found in medical students following 1 month of academic stress (Liston, McEwen, & Casey, 2009). Similarly, in animal models, stress produces deficits on a variety of cognitive and emotion-regulation tasks, including extinction of conditioned fear, attentional set-shifting, spatial learning and recognition, and working memory (reviewed in Conrad, Ortiz, & Judd, 2017; Hurtubise & Howland, 2017; Maren & Holmes, 2016; Wellman & Moench, 2019). Interestingly, many of the behaviors that are susceptible to stress are mediated by the prefrontal cortex.

As in primates, the prefrontal cortex in rodents can be subdivided into several major subregions (Fig. 1). The medial prefrontal cortex includes anterior cingulate, prelimbic, and infralimbic cortex. This region is functionally homologous to the primate dorsolateral and ventromedial prefrontal cortices, and plays a role in autonomic and HPA axis regulation (McKlveen, Myers, & Herman, 2015), emotion regulation (Wellman & Moench, 2019), working memory (Kesner & Churchwell, 2011), and cognitive flexibility (Hamilton & Brigman, 2015). Orbitofrontal cortex, which includes the

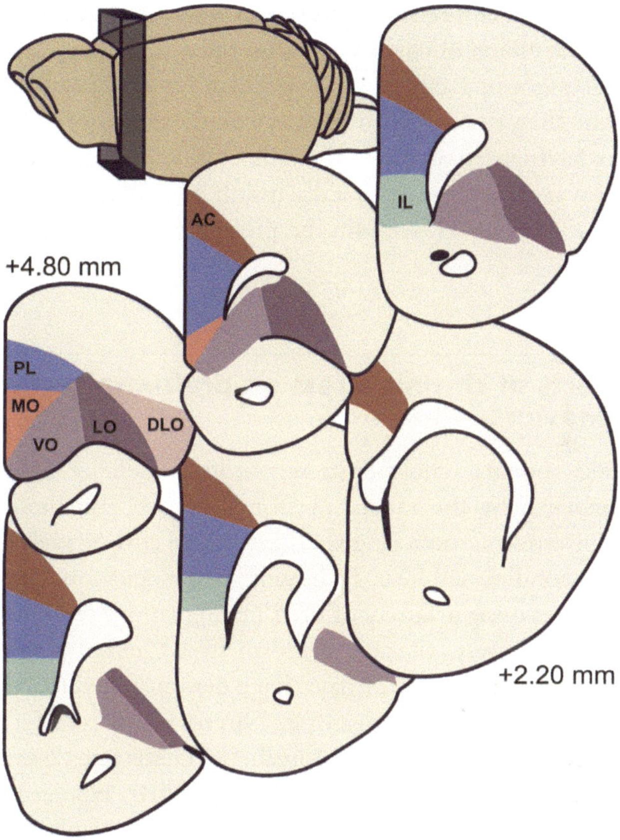

Fig. 1 Schematic diagram of coronal hemisections through the rostral forebrain of the rat, with prefrontal brain regions likely to play a role in stress effects on extinction identified. Coordinates given are relative to Bregma in rat brain. AC, anterior cingulate; PL, prelimbic; IL, infralimbic; MO, medial orbitofrontal; VO, ventral orbitofrontal; LO, lateral orbitofrontal. *Adapted from Paxinos, G., & Watson, C. (1998). The rat brain in stereotaxic coordinates (4th ed.). New York: Academic Press.*

medial, ventral, and lateral orbitofrontal subregions, is functionally homologous to primate orbitofrontal cortex and appears to play a role in modulating behavioral responses based on changing incentive values of reward-related stimuli (Hamilton & Brigman, 2015; Holmes & Wellman, 2009).

While stress influences the structure and function of orbitofrontal cortex (e.g., Godar et al., 2015; Gourley, Swanson, & Koleske, 2013; Liston et al., 2006), these effects have been most extensively studied in the medial prefrontal cortex. Thus, in this chapter we focus on the medial prefrontal cortex. First, we provide a brief overview of the effects of chronic stress on behaviors mediated by the medial prefrontal cortex, as assessed in animal models. We then focus on the effects of chronic stress on the morphology and physiology of the medial prefrontal cortex, and mechanisms that may mediate these effects. Because the vast majority of research on the effects of stress on the prefrontal cortex has focused on males, the bulk of this chapter will likewise focus on males. However, we end with a discussion of emerging evidence for sex differences in the effects of stress on the prefrontal cortex.

2. Effects of chronic stress on prefrontally-mediated behaviors

In male rodents, chronic stress produces deficits in a variety of behaviors mediated by the medial prefrontal cortex, and these deficits are remarkably consistent across a range of stressors and chronicities. For instance, extinction of conditioned fear is heavily dependent on the infralimbic cortex, and studies using a variety of stress manipulations (e.g., restraint stress, unpredictable mild stress, elevated platform stress), either acute or chronic (varying in duration from 7 to 21 days) have demonstrated that stress impairs extinction learning (Baran, Armstrong, Niren, Hanna, & Conrad, 2009; Garcia, Spennato, Nilsson-Todd, Moreau, & Deschaux, 2008; Goswami, Cascardi, Rodriguez-Sierra, Duvarci, & Pare, 2010; Izquierdo, Wellman, & Holmes, 2006; Knox, George, et al., 2012; Knox, Nault, Henderson, & Liberzon, 2012; Knox, Perrine, George, Galloway, & Liberzon, 2010; Maroun et al., 2013; Miracle, Brace, Huyck, Singler, & Wellman, 2006; Wilber et al., 2011). Likewise, chronic stress of varying intensities and chronicities impairs behavioral flexibility as assessed via attentional set-shifting. The extradimensional shift in this task is mediated by the prelimbic cortex (Birrell & Brown, 2000; Hamilton & Brigman, 2015), and is preferentially disrupted by chronic stress in male rats (Jett, Bulin, Hatherall, McCartney, & Morilak, 2017; Liston et al., 2006; McKlveen et al., 2016; Moench, Breach, & Wellman, 2020; Nikiforuk & Popik, 2011, 2013). Likewise, working memory, which is also dependent prelimbic function, is impaired after chronic stress (Hains et al., 2009; Kim et al., 2017; Mika et al., 2012; Mizoguchi et al., 2000). On the other hand, effects of acute stress on working memory are mixed, with reports of either facilitation (Yuen et al., 2009) or impairment (Devilbiss, Jenison, & Berridge, 2012), perhaps dependent on the timing of the acute stressor.

3. Effects of chronic stress on structure and function of the prefrontal cortex

3.1 Effects of chronic stress on the medial prefrontal cortex: Morphology

The medial prefrontal cortex is a target for hormones involved in the stress response, such as the glucocorticoids corticosterone and cortisol (Meaney & Aitken, 1985). Stress-induced alterations in neuronal morphology are perhaps the best-documented effects of stress on the prefrontal cortex (Fig. 2). In male rats, chronic restraint stress produces retraction of apical dendrites of pyramidal neurons in the prelimbic cortex (Cook & Wellman, 2004; Garrett & Wellman, 2009; Liston et al., 2006; Martin & Wellman, 2011; Moench & Wellman, 2017; Radley et al., 2004, 2005, 2006), an effect that

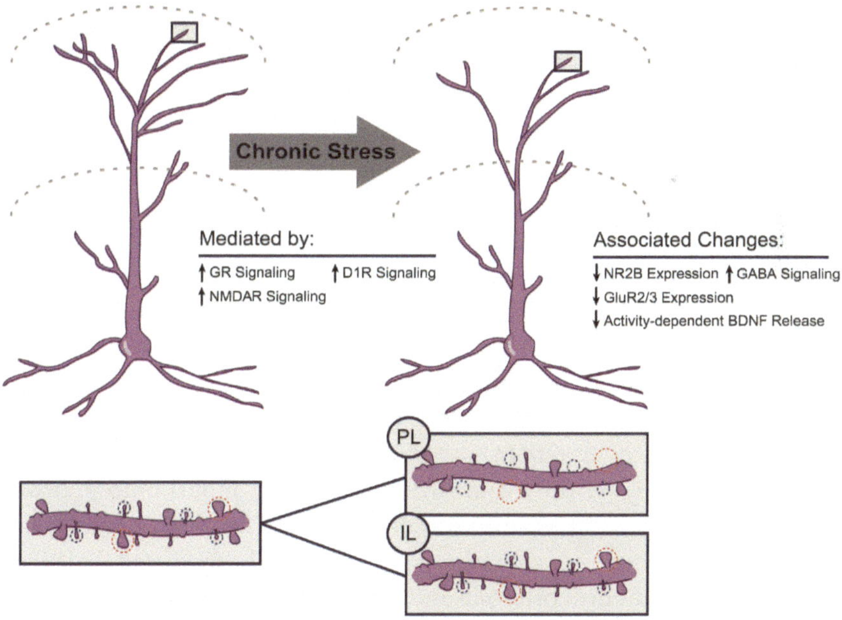

Fig. 2 Summary of chronic stress-induced neuronal remodeling in medial prefrontal cortex of male rodents. (Top) Chronic stress produces retraction of apical dendrites of pyramidal neurons in both prelimbic and infralimbic cortex. Blockade of either glucocorticoid (GR), NMDA (NMDAR), or dopaminergic D1 family receptors prevents the stress-induced dendritic remodeling. (Bottom) Chronic stressors decrease spine density in prelimbic cortex and the ratio of mushroom (circled in blue, large heads) to thin spines (circled in red, small heads). The effect of stress on spine density in infralimbic cortex is less studied, but may be less robust.

is mimicked with chronic corticosterone administration (Cerqueira et al., 2005; Cerqueira, Taipa, Uylings, Almeida, & Sousa, 2007; Wellman, 2001). A similar pattern of stress-induced retraction is seen in apical dendritic branches of neurons within the infralimbic region of the medial prefrontal cortex (Izquierdo et al., 2006; Moench, Maroun, Kavushansky, & Wellman, 2016; Shansky, Hamo, Hof, McEwen, & Morrison, 2009). Even shorter, milder episodes of stress are sufficient to produce dendritic atrophy in the medial prefrontal cortex: 10 min of restraint stress for 10 days (Brown, Henning, & Wellman, 2005) or 3 weeks of vehicle injection alone (Wellman, 2001) reduce dendritic arborization within the medial prefrontal cortex, again with retraction occurring only in distal portions of the apical arbor. These findings suggest that the morphology of the medial prefrontal cortex is exquisitely sensitive to stress. Indeed, even acute stress, such as a single episode of forced swimming in mice (Izquierdo et al., 2006) or exposure to an elevated platform stressor in rats (Moench et al., 2016) produces reductions in apical dendritic branch length in the infralimbic cortex.

Variations in dendritic structure are critical determinants of neuronal activity (e.g., Rall et al., 1992; Spruston, 2008). In addition, within a neuron, dendritic compartments (e.g., apical vs basilar, distal tufts vs proximal trunk; Gordon, Polsky, & Schiller, 2006; Han & Heinemann, 2013) are functionally distinct, with different inputs across dendritic compartments (e.g., Groenewegen, 1988; Swanson & Cowan, 1977; Urban-Ciecko & Barth, 2016). Thus, stress-induced structural alterations, with their specificity to the apical dendrite and their distinctive spatial distribution within the apical arbor, may have important functional implications. Consistent with this notion, in stressed rats, apical dendritic debranching and retraction in the infralimbic region of medial prefrontal cortex predict poorer extinction memory (Moench et al., 2016).

Dendritic spines are the major sites of excitatory inputs onto prefrontal pyramidal cells; thus, stress-induced alterations in dendritic spine density could also contribute to alterations in prefrontally-mediated behaviors. Indeed, stress- and corticosterone-induced dendritic retraction is coupled with a decrease in spine density in the anterior cingulate and prelimbic cortices (Barfield et al., 2017; Hains et al., 2009; Liu & Aghajanian, 2008; Radley et al., 2006, 2008; Swanson, Shapiro, Whyte, & Gourley, 2013). Importantly, this decrease in spine density correlates with impairment of working memory, and prevention of the stress-induced spine loss via inhibition of protein kinase C prevents the working memory deficit (Hains et al., 2009).

Variations in spine morphology may also reflect differences in function. For instance, larger, mushroom-type spines are perhaps more mature, less labile, so-called "memory spines," whereas thin spines are thought to be more plastic (Segal, 2017). Accordingly, numerous alterations in dendritic spine morphology after acute or chronic stress have been reported. For instance, a shift in spine morphology from large mushroom spines to smaller thin spines, and thus an overall reduction in spine volume and surface area, has been demonstrated in the dorsal medial prefrontal cortex following either 3 weeks of daily restraint stress (Barfield et al., 2017; Radley et al., 2008; though cf. effects of chronic corticosterone administration, Seib & Wellman, 2003) or 2 weeks of chronic variable stress (Radley, Anderson, Hamilton, Alcock, & Romig-Martin, 2013).

On the other hand, the effects of stress and glucocorticoids on dendritic spines in infralimbic cortex are less clear. In mice, administration of corticosterone via drinking water at a dose that produces restraint stress levels of corticosterone in plasma significantly reduced dendritic spine density on apical branches of deep-layer pyramidal cells in the infralimbic cortex (Gourley et al., 2013). However, chronic restraint stress (6 h/day for 3 weeks) did not significantly alter spine density in the infralimbic cortex of rats (Shansky et al., 2009). These disparate results may reflect sustained high levels of corticosterone in the former study vs habituation to restraint in the latter, which suggests that subregions of the medial prefrontal cortex may be differentially sensitive to stress. However, alterations in spine morphology or density accompanying sub-chronic or acute stress may nonetheless contribute to stress-induced deficits in extinction, as density of thin spines and estimates of total numbers of spines on apical terminal branches are strongly negatively correlated with freezing during extinction in rats exposed to one episode of elevated platform stress (Moench et al., 2016). Important areas for future research include understanding how subregion-specific stress-induced neuronal remodeling influences the physiology of the medial prefrontal cortex and related circuitry and how these influences are expressed in prefrontally-mediated behaviors.

3.2 Effects of chronic stress on the medial prefrontal cortex: Physiology

Stress-induced changes in the activity of the medial prefrontal cortex may underlie deficits in prefrontally-mediated behaviors. For instance, in male rats, the deficit in extinction retrieval induced by 7 days of prior chronic restraint stress is accompanied by alterations in activity of infralimbic cortex neurons (Wilber et al., 2011): neurons in unstressed rats exhibited increased firing in response to the conditioned stimulus, whereas in stressed rats, this increase in firing was absent. In addition, chronic restraint stress impaired induction of long-term potentiation in the hippocampus-to-medial prefrontal cortex pathway (Cerqueira, Mailliet, Almeida, Jay, & Sousa, 2007). Given that hippocampal projections to the prefrontal cortex play a critical role in extinction learning (Garcia et al., 2008; Knapska et al., 2012; Knapska & Maren, 2009), this impaired plasticity could contribute to the stress-induced impairment of extinction (Garcia et al., 2008). Similarly, even an acute elevated platform stressor impairs the induction of long-term potentiation in the amygdala-to-medial prefrontal cortex pathway (Maroun & Richter-Levin, 2003), and application of the Single Prolonged Stressor model decreases functional connectivity of the ventral medial prefrontal cortex and the basolateral amygdala, as assessed with expression of the immediate early gene c-Jun during extinction retrieval (Knox et al., 2018). Finally, acute stress exposure during a working memory task altered the firing of prelimbic neurons, and this altered activity was associated with deficits on the task (Devilbiss et al., 2012). Taken together, these studies provide support for the notion that stress-induced alterations in the medial prefrontal cortex contribute to deficits in prefrontally-mediated behaviors.

3.3 Effects of chronic stress on medial prefrontal cortex: Neurochemistry

Stress-induced changes in neuronal morphology and function are accompanied by pronounced alterations in the prefrontal glutamatergic system. For instance, 2 weeks of exposure to stress levels of corticosterone administered via drinking water markedly decreased expression of the NR2B subunit of the NMDA receptor and the GluR2 and 3 subunits of the AMPA receptor in the ventral medial prefrontal cortex (comprised of infralimbic cortex and ventral portions of prelimbic cortex), and this was associated with impaired extinction of contextual fear (Gourley, Kedves, Olausson, & Taylor, 2009). Likewise, 7 days of daily restraint or variable stress altered expression of the NR1, NR2A, and NR2B subunits of the NMDA receptor and the GluR1 and GluR1 subunits of the AMPA receptor in adolescent (1-month-old) rats, and impaired AMPA- and NMDA receptor-mediated neuronal excitability in the medial prefrontal cortex (Wei et al., 2014; Yuen et al., 2012). Similar stress-induced decreases in glutamatergic receptor expression and transmission have been reported in the medial prefrontal cortex of adult rats and mice (Jett et al., 2017; Shepard & Coutellier, 2018). Given the well-documented role for NMDA receptor-dependent activity in the prefrontal cortex in prefrontally-mediated behaviors such as extinction (Burgos-Robles, Vidal-Gonzalez, Santini, & Quirk, 2007; Santini, Muller, & Quirk, 2001; Vieira et al., 2015), it is likely that this stress-induced dysfunction of the prefrontal glutamatergic system contributes to the stress-induced impairment of prefrontal behaviors.

Interestingly, changes in the glutamatergic system may be driven in part by stress-induced increases in GABAergic inhibition of pyramidal cells. Using patch-clamp recordings, Herman and colleagues (McKlveen et al., 2016) recently demonstrated that 2 weeks of chronic variable stress (rotation on a platform orbital shaker, warm swim, cold swim, cold exposure, brief hypoxia, overnight social isolation, overnight social crowding) produced increased miniature inhibitory postsynaptic currents in pyramidal neurons in infralimbic cortex, an effect that was blocked by the application of a $GABA_A$ receptor antagonist. This increased inhibition may have been due to increased GABAergic innervation of the pyramidal cells, as the number of Gad65-positive puncta on CAMKII-positive cells was markedly increased in rats subjected to chronic variable stress (McKlveen et al., 2016).

4. Long-term sequelae of chronic stress

Recent data suggest that chronic stress can have relatively long-lasting effects on morphology and neurochemistry in medial prefrontal cortex. For instance, dendritic remodeling in prelimbic cortex of chronically stressed male rats is highly dynamic during the post-stress rest period. Following a 7-day post-chronic stress rest period, chronically stressed male rats have apical dendritic outgrowth relative to unstressed males, despite initial chronic stress-induced dendritic retraction (Moench & Wellman, 2017). In this same time period, expression of the genes coding for the NR1 and GluR1 subunits of the NMDA and AMPA receptors, respectively, along with Gad67 and parvalbumin, is increased (Moench et al., 2020). It is interesting to speculate that the upregulation of glutamatergic genes may reflect the increase in dendritic material seen 1 week post-stress, while the upregulation of the GABAergic genes may be indicative of compensatory changes involved in maintaining an optimal excitation-inhibition balance in prelimbic cortex. Interestingly, this same study demonstrated a chronic stress-induced deficit in extradimensional set-shifting 1 day following stress, which was no longer present 1 week post-stress. This suggests that the delayed changes in gene expression may also play a role in the functional recovery of prelimbic cortex following the cessation of chronic stress (Moench et al., 2020). These lasting changes may confer some resilience to subsequent stressors, as chronically stressed male rats have persistent reductions in novel stress-induced c-Fos expression in mPFC (Moench, Breach, & Wellman, 2019; Ostrander et al., 2009) and a blunted neuroendocrine response to a novel stress challenge (Ostrander et al., 2009).

5. Mechanisms of chronic stress-induced alterations in the medial prefrontal cortex

A growing body of literature has begun to elucidate the mechanisms underlying stress-induced changes in structure and function in the medial prefrontal cortex. For instance, similar to stress, 3 weeks of daily injections of stress levels of corticosterone remodels apical dendrites of pyramidal neurons in the dorsal medial prefrontal cortex (Cerqueira et al., 2005; Wellman, 2001). Further, systemic administration of the glucocorticoid receptor blocker RU38486 during either chronic restraint (10 days) or chronic unpredictable stress (14 days) prevents stress-induced apical dendritic remodeling (Liu & Aghajanian, 2008) and spine loss (Horchar & Wohleb, 2019) in the medial prefrontal cortex. This is consistent with the finding that, in vitro, glucocorticoid receptors mediate the stress-induced decreases in functional glutamatergic receptors and impairment of glutamatergic transmission in medial prefrontal cortex (Yuen et al., 2012). However, it is not known whether, in vivo, glucocorticoid receptors contribute to stress-induced dendritic retraction via direct actions in prefrontal cortex, or by modulation of inputs to prefrontal cortex.

In addition, systemic administration of a competitive NMDA receptor blocker during daily restraint stress also prevents stress-induced dendritic retraction in medial prefrontal cortex (Martin & Wellman, 2011). Again, this could be due either to a direct effect in the medial prefrontal cortex or alteration of inputs to the medial prefrontal cortex. Although the aforementioned study did not assess spine density, a small literature implicates NMDARs and AMPA receptors (AMPARs) in chronic stress-induced loss of spines. For instance, a single dose of either the NMDAR blocker ketamine or the specific NMDAR 2B blocker Ro25-6981 following 21 days of chronic unpredictable mild stress (including cold, disruption of light-dark cycle, crowding, shaking, and exposure to an aversive odor) rescues spines, particularly in layer V of the prelimbic cortex (Li et al., 2011). This effect involved activation of the mTOR (mammalian target of rapamaycin) signaling pathway, as the beneficial effect of NMDAR blockade was prevented by administration of rapamycin. Additionally, NMDAR- and AMPAR-excitatory postsynaptic currents (EPSCs) are reduced following 5 or more days, but not 3 or fewer days, of repeated behavioral stressors (Li et al., 2011; Yuen et al., 2012). Further examination of the mechanisms underlying decreased activity at these glutamatergic receptors demonstrated that the reduction in EPSC amplitude is dependent on ubiquitin/proteasome degradation of GluR1 and NR1 subunits (Yuen et al., 2012). Given the profound and multifaceted influence of glucocorticoids and glucocorticoid receptors on glutamatergic transmission, including both classical and nongenomic effects on glutamate release and clearance as well as NMDAR and AMPAR responses (reviewed in Popoli, Yan, McEwen, & Sanacora, 2012), glutamatergic receptors on dendritic spines might be the final common pathway by which stress-induced increases in glucocorticoids alter both dendritic morphology in the medial prefrontal cortex and prefrontally-mediated behaviors.

Consistent with this notion, modulators of glutamatergic activity, such as dopaminergic signaling and brain-derived neurotrophic factor (BDNF), are critical for several prefrontally-mediated behaviors and have also been implicated in stress-induced alterations in the medial prefrontal cortex. For instance, dopamine modulates NMDA receptor-mediated currents via D1 receptors (reviewed in Tritsch & Sabatini, 2012), and D1 receptors in the medial prefrontal cortex have been implicated in extinction (Fiorenza, Rosa, Izquierdo, & Myskiw, 2012; Hikind & Maroun, 2008), attentional set-shifting (Nikiforuk, 2012), and working memory (Vijayraghavan, Wang, Birnbaum, Williams, & Arnsten, 2007). Given the interactions between glucocorticoids and D1 receptors (reviewed in Sinclair, Purves-Tyson, Allen, & Weickert, 2014), it is not surprising that infusion of the D1 blocker SCH23390 into the medial prefrontal cortex prevents stress-induced dendritic retraction (Lin, Borders, Lundewall, & Wellman, 2015) and attenuates stress-facilitated, priming-induced drug seeking after extinction (Ball et al., 2018).

BDNF signaling is likewise critical for extinction (McGinty, Whitfield, & Berglind, 2010; Peters, Dieppa-Perea, Melendez, & Quirk, 2010; Rosas-Vidal, Do-Monte, Sotres-Bayon, & Quirk, 2014) and behavioral flexibility (Barfield & Gourley, 2017; Otis, Fitzgerald, & Mueller, 2014; Papaleo et al., 2011). However, although BDNF appears to play a key role in synaptic remodeling in hippocampus (An et al., 2008) and amygdala (Govindarajan et al., 2006) following stress, evidence for its role in the medial prefrontal cortex is mixed, with some reports of stress-induced downregulation of BDNF mRNA (reviewed in Calabrese, Molteni, Racagni, & Riva, 2009), and other studies failing to find stress-induced changes (Chiba et al., 2012; Lin et al., 2009). On the other hand, infusion of BDNF into the infralimbic cortex can reverse some of the behavioral effects of stress (Graybeal et al., 2011). Further, Yu et al. (2012) demonstrated that 7 days of chronic restraint stress reduced spine density in the medial prefrontal cortex of Val66Met knock-in mice, in which activity-dependent release of BDNF is reduced, compared to stressed wild type (WT) mice. Interestingly, although the behavioral and neuroendocrine profile of these knock-in mice was comparable to that of WT prior to stress exposure, spine density was not different in unstressed WT vs knock-in mice. Nonetheless, stress-induced increases in plasma corticosterone and adrenocorticotropic hormone (ACTH) were greater in knock-in mice relative to WT. Thus, low basal levels of BDNF are not sufficient to decrease spine density, but could contribute to the development of a heightened response to stress, leading to downstream effects on spinogenesis. This finding illustrates the importance of considering gene × environment interactions in the pathophysiology of stress-sensitive disorders. In addition, administration of ketamine, an NMDA antagonist, did not ameliorate the effects of BDNF deficiency in Val66Met knock-in mice, but does ameliorate stress-induced prefrontal spine loss in WT mice and rats, indicating that BDNF plays a crucial role in synaptogenesis in the medial prefrontal cortex. Liu et al. (2012) suggest a mechanism by which glucocorticoids, NMDARs, and BDNF interact to mediate synaptogenesis: (1) ketamine transiently increases the presynaptic release of glutamate, leading to a burst of action potentials at the synapse; (2) this bursting increases AMPAR stimulation, facilitating the release of BDNF; and (3) BDNF binds to its receptor (TrkB), activating mTOR, which has the downstream effect of the translation of synaptic proteins, thus leading to synaptogenesis (Liu et al., 2012). Therefore, if BDNF is down-regulated as in Val66Met knock-in mice or by chronic stress, the downstream events of ketamine administration or natural rises in glutamate will be blocked, resulting in either synaptic stasis or potentially, synaptic pruning in cases where excess levels of glucocorticoids are present (see Fig. 3). This synaptic pruning in turn could contribute to stress-induced deficits in extinction.

Interestingly, emerging evidence suggests a novel potential mechanism for the effects of stress on prefrontal structure and function in males: activation of microglia. Microglia, the immunocompetent cells of the brain, can

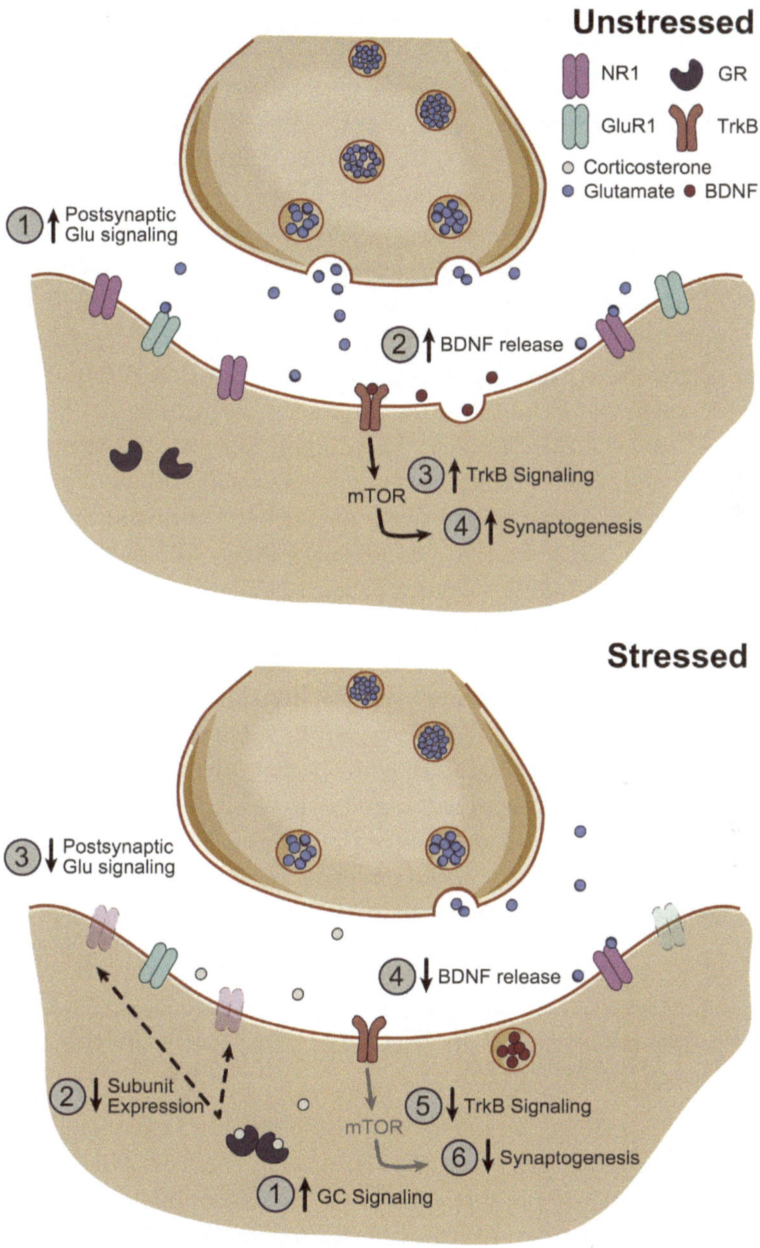

Fig. 3 Possible neurochemical/molecular pathways for stress-induced synaptic remodeling in medial prefrontal cortex in males. (Top) In an unstressed rodent, increases in postsynaptic glutamatergic signaling (1) are associated with release of brain-derived neurotrophic factor (BDNF) (2); BDNF's activity at its TrkB receptor (3) results in increased synaptogenesis via a mammalian target of rapamaycin (mTOR)-dependent pathway (4). (Bottom) During chronic stress, increased glucocorticoid signaling (1) results in loss of functional NMDA and AMPA receptors (2), which decreases postsynaptic glutamatergic signaling (3). Concurrent reduction in BDNF release (4) in medial prefrontal cortex decreases TrkB signaling (5), resulting in decreased synaptogenesis (6). NR1, NR1 subunit of the NMDA receptor; GluR1, GluR1 subunit of the AMPA receptor; GR, glucocorticoid receptor.

sculpt neuronal morphology through the release of neurotrophic factors, cytokines, and chemokines, engagement with complement proteins, and direct microglia-neuron contact. These cells can modulate neuronal activity, prune synapses, stimulate spine outgrowth, and alter behavioral function (reviewed in Salter & Beggs, 2014; Salter & Stevens, 2017). Chronic stress increases microglial cell density and heightens microglia-neuron interaction, as reflected in cellular morphology and reciprocal increases in neuronal colony stimulating factor 1 (CSF1) and microglial CSF1R expression in the medial prefrontal cortex (Bollinger, Bergeon Burns, & Wellman, 2016; Bollinger, Salinas, Fender, Sengelaub, & Wellman, 2019; Walker, Nilsson, & Jones, 2013; Wohleb, Terwilliger, Duman, & Duman, 2018). These changes in microglial activation are associated with deficits in spatial working memory and temporal object recognition (Hinwood, Morandini, Day, & Walker, 2012; Horchar & Wohleb, 2019; Wohleb et al., 2018). Moreover, recent findings indicate that chronic unpredictable stress increases microglial engagement with dendritic spines and phagocytosis of dendritic elements, while decreasing apical dendritic spine density in the medial prefrontal cortex. Importantly, local knockdown of CSF1 prevented these effects (Wohleb et al., 2018), suggesting that neuron-microglia signaling may orchestrate stress-induced dendritic remodeling. In line with prior findings, glucocorticoids appear to initiate these interactions: blocking glucocorticoid receptors with daily injections of RU38486 prevents chronic stress-induced, microglia-mediated dendritic remodeling in the medial prefrontal cortex (Horchar & Wohleb, 2019).

6. Sex differences in the effect of stress on the medial prefrontal cortex

Despite dramatic sex differences in the rates and expression of stress-related psychological disorders (Cover, Maeng, Lebrón-Milad, & Milad, 2014), the vast majority of the research on the neurobiological mechanisms underlying stress effects on emotional behavior has focused on males, as have the preceding sections of this chapter. Investigations of the mechanisms underlying potential stress-induced plasticity of corticolimbic structures in females are critical for providing the groundwork necessary to develop sex-specific treatment for stress-related psychopathology. A small number of studies have begun to address this issue, and have demonstrated that stress does in fact produce very different effects on the structure and function of

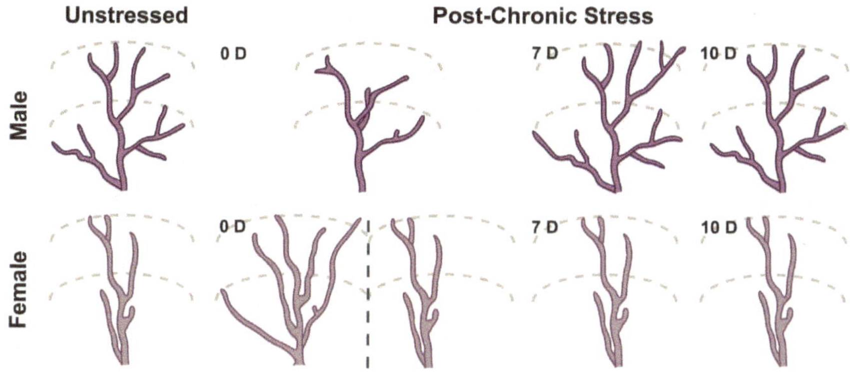

Fig. 4 Sex differences in stress effects on dendritic morphology and recovery in the medial prefrontal cortex. Pyramidal neurons in the medial prefrontal cortex have more complex apical dendritic arbors in unstressed male rats compared to unstressed females. Following 10 days of chronic restraint stress (0 D), males show robust reductions in dendritic arborization, whereas females show either no change in arborization or dendritic outgrowth. Males exhibit dendritic outgrowth at 7 days post-stress (7 D), with dendritic complexity returning to baseline by 10 days post-stress (10 D). Females show baseline dendritic complexity at both 7- and 10-days post-stress. Together, these findings indicate sex differences in dendritic arborization in the medial prefrontal cortex, alongside sex-specific stress effects on dendritic complexity in this region. The physiological and behavioral consequences of these sex-dependent patterns remain largely unexplored.

the prefrontal cortex. For example, while chronic stress impairs temporal object recognition and fear extinction in male rats, similar deficits are not observed in female rats (Baran et al., 2009; Hoffman, Armstrong, Hanna, & Conrad, 2010; Wei et al., 2014). Likewise, chronic stress impairs attentional set-shifting in male but not female rats (Moench et al., 2020). Further, whereas male rats exhibit dendritic retraction in medial prefrontal cortex following stress, female rats either exhibit no dendritic changes or may even exhibit dendritic hypertrophy (Fig. 4; Garrett & Wellman, 2009; Moench & Wellman, 2017; Shansky et al., 2010).

In males, chronic stress results in a downregulation of the NMDA receptor subunits NR1, NR2A, and NR2B (Lee & Goto, 2011; Shepard & Coutellier, 2018). These changes are associated with deficits in temporal order working memory in adolescent rats (Wei et al., 2014). Downregulation of the AMPA receptor subunit GluR1 also has also been reported following chronic stress and is associated with depressive-like behaviors in male rats (Li et al., 2011). Thus, reduced glutamatergic transmission likely plays an important role in stress-induced deficits in behaviors that are mediated by the medial prefrontal cortex. Notably, the effect of stress on glutamatergic neurotransmission appears to be more pronounced in males than females (Wei et al., 2014).

In contrast to the effects of chronic stress on glutamatergic neurotransmission, recent studies have suggested that chronic stress-induced changes in GABAergic neurotransmission may be more pronounced in the medial prefrontal cortex of females. For instance, female mice exposed to 2 weeks of chronic unpredictable mild stress have an increase in the expression of parvalbumin mRNA in the medial prefrontal cortex, which corresponded to a decrease in overall neuronal activation as measured by c-Fos expression (Shepard, Page, & Coutellier, 2016). Further, chronically stressed female mice have greater glutamatergic transmission onto parvalbumin-expressing neurons in the medial prefrontal cortex (Shepard & Coutellier, 2018). Thus, inhibitory tone in the medial prefrontal cortex may be enhanced in chronically stressed female mice, which may contribute to deficits on an object-context mismatch test (Shepard et al., 2016; Shepard & Coutellier, 2018). Together, these studies suggest that chronic stress-induced changes in the prefrontal GABAergic system may be pronounced in females, while changes in glutamatergic neurotransmission may be pronounced in males. This pattern likely contributes to sex differences in performance on tasks mediated by the medial prefrontal cortex following chronic stress.

Differential activation of microglia may contribute to these sex-dependent effects of stress on the prefrontal cortex and behavior. For instance, acute restraint stress-induced microglial morphological activation in medial prefrontal cortex in males, but deactivation in females (Bollinger et al., 2016). Microglial morphology returned to baseline following 10 days of restraint in males, whereas microglial deactivation persisted in females (Bollinger et al., 2016). These findings demonstrate both opposite effects of stress on microglia activation in males and females and differing temporal patterns of stress-induced microglial remodeling. Consistent with the notion that these differences may contribute to sex-dependent stress effects on neuronal morphology, chronic unpredictable stress-induced microglia-neuron interaction and synaptic pruning in medial prefrontal cortex in males, but not females (Wohleb et al., 2018).

Interestingly, the lasting effects of chronic stress are also sex-dependent. For instance, while male rats undergo dynamic dendritic remodeling in the week following the cessation of chronic restraint stress, females do not (Moench & Wellman, 2017). Such differences may produce functional differences in subsequent responses to stress. For example, there are sex differences in neuronal activation across a number of corticolimbic brain regions in chronically stressed rats that are exposed to a novel stress challenge (Moench et al., 2019). Notably, males have a persistent reduction in novel stress-induced neuronal activation in prelimbic cortex. In contrast, prior chronic stress does not modulate neuronal activation in prelimbic cortex in response to a novel stress challenge in females (Moench et al., 2019). Consistent with this, while parvalbumin mRNA expression was increased in males 1 week post-chronic stress and sustained through a novel stress challenge, this novel stress challenge reduced parvalbumin expression in females. Together, these data suggest that males may have enhanced inhibitory signaling in prelimbic cortex following a novel stress challenge that was not present immediately following chronic stress. In contrast, while females appear to have greater inhibitory tone after chronic stress, this inhibition may be reduced following a novel acute stress challenge, resulting in behavioral deficits that are not found in males (Moench et al., 2020). These differential responses in the medial prefrontal cortex coincide with sex-dependent effects of the subsequent novel stressor on attentional set-shifting, with females but not males showing deficits in extradimensional set-shifting (Moench et al., 2020).

Further systematic investigation of the phenomenology, functional implications, and mechanisms underlying the immediate and longer-term sex-dependent effects of stress is critical for developing sex-specific treatments for stress-sensitive psychopathologies.

7. Conclusions

Extensive evidence gathered from animal models demonstrates that prior stress—either chronic or acute—alters the prefrontal cortex and behaviors mediated by it. Indeed, in males, these behavioral changes correlate nicely with alterations in prefrontal structure and function. However, much work remains to be done, both at the level of neurochemical mechanisms and localization of these mechanisms to specific structures, to directly test the hypothesis that these changes mediate stress-induced changes in prefrontally-dependent behaviors. Ultimately, a full understanding of the neural mechanisms underlying stress-induced impairments in prefrontal function will require investigations at the circuit level of analysis, to account for differential and opposing effects of stress on the neural circuitry underlying these complex behaviors (Wellman & Moench, 2019). Finally, investigation of the neural basis for individual differences, especially sex differences in vulnerability to stress-induced alterations in extinction and behavioral flexibility, will be instrumental in identifying avenues for novel interventions for stress-sensitive disorders characterized by alterations in prefrontally-mediated behaviors.

Conflict of interest statement

The authors declare that they have no conflict of interest.

References

An, J. J., Gharami, K., Liao, G.-Y., Woo, N. H., Lau, A. G., Vanevski, F., et al. (2008). Distinct role of long 3' UTR BDNF mRNA in spine morphology and synaptic plasticity in hippocampal neurons. *Cell, 134*(1), 175–187.

Ball, K. T., Stone, E., Best, O., Collins, T., Edson, H., Hagan, E., et al. (2018). Chronic restraint stress during withdrawal increases vulnerability to drug priming-induced cocaine seeking via a dopamine D1-like receptor-mediated mechanism. *Drug and Alcohol Dependence, 187*, 327–334. https://doi.org/10.1016/j.drugalcdep.2018.03.024.

Baran, S. E., Armstrong, C. E., Niren, D. C., Hanna, J. J., & Conrad, C. D. (2009). Chronic stress and sex differences on the recall of fear conditioning and extinction. *Neurobiology of Learning and Memory, 91*(3), 323–332.

Barfield, E. T., Gerber, K. J., Zimmermann, K. S., Ressler, K. J., Parsons, R. G., & Gourley, S. L. (2017). Regulation of actions and habits by ventral hippocampal trkB and adolescent corticosteroid exposure. *PLoS Biology, 15*(11), e2003000. https://doi.org/10.1371/journal.pbio.2003000.

Barfield, E. T., & Gourley, S. L. (2017). Adolescent corticosterone and TrkB pharmaco-manipulations Sex-dependently impact instrumental reversal learning later in life. *Frontiers in Behavioral Neuroscience, 11*, 237. https://doi.org/10.3389/fnbeh.2017.00237.

Birrell, J. M., & Brown, V. J. (2000). Medial frontal cortex mediates perceptual attentional set shifting in the rat. *The Journal of Neuroscience, 20*(11), 4320–4324.

Bollinger, J. L., Bergeon Burns, C. M., & Wellman, C. L. (2016). Differential effects of stress on microglial cell activation in male and female medial prefrontal cortex. *Brain, Behavior, and Immunity, 52*, 88–97.

Bollinger, J. L., Salinas, I., Fender, E., Sengelaub, D. R., & Wellman, C. L. (2019). Gonadal hormones differentially regulate sex-specific stress effects on glia in the medial prefrontal cortex. *Journal of Neuroendocrinology, 31*(8). e12762. https://doi.org/10.1111/jne.12762.

Brown, S., Henning, S., & Wellman, C. L. (2005). Short-term, mild stress alters dendritic morphology in rat medial prefrontal cortex. *Cerebral Cortex, 15*, 1714–1722.

Burgos-Robles, A., Vidal-Gonzalez, I., Santini, E., & Quirk, G. J. (2007). Consolidation of fear extinction requires NMDA receptor-dependent bursting in the ventromedial prefrontal cortex. *Neuron, 53*(6), 871–880.

Calabrese, F., Molteni, R., Racagni, G., & Riva, M. A. (2009). Neuronal plasticity: A link between stress and mood disorders. *Psychoneuroendocrinology, 34*(Suppl. 1), S208–S216. https://doi.org/10.1016/j.psyneuen.2009.05.014.

Cerqueira, J. J., Mailliet, F., Almeida, O. F. X., Jay, T. M., & Sousa, N. (2007). The prefrontal cortex as a key target of the maladaptive response to stress. *Journal of Neuroscience, 27*(11), 2781–2787. https://doi.org/10.1523/jneurosci.4372-06.2007.

Cerqueira, J. J., Pego, J. M., Taipa, R., Bessa, J. M., Almeida, O. F. X., & Sousa, N. (2005). Morphological correlates of corticosteroid-induced changes in prefrontal cortex-dependent behaviors. *Journal of Neuroscience, 25*(34), 7792–7800. https://doi.org/10.1523/jneurosci.1598-05.2005.

Cerqueira, J. J., Taipa, R., Uylings, H. B. M., Almeida, O. F. X., & Sousa, N. (2007). Specific configuration of dendritic degeneration in pyramidal neurons of the medial prefrontal cortex induced by differing corticosteroid regimens. *Cerebral Cortex*, *17*(9), 1998–2006.

Chiba, S., Numakawa, T., Ninomiya, M., Richards, M. C., Wakabayashi, C., & Kunugi, H. (2012). Chronic restraint stress causes anxiety- and depression-like behaviors, downregulates glucocorticoid receptor expression, and attenuates glutamate release induced by brain-derived neurotrophic factor in the prefrontal cortex. *Progress in Neuro-Psychopharmacology & Biological Psychiatry*, *39*, 112–119.

Conrad, C. D., Ortiz, J. B., & Judd, J. M. (2017). Chronic stress and hippocampal dendritic complexity: Methodological and functional considerations. *Physiology & Behavior*, *178*, 66–81. https://doi.org/10.1016/j.physbeh.2016.11.017.

Cook, S. C., & Wellman, C. L. (2004). Chronic stress alters dendritic morphology in rat medial prefrontal cortex. *Journal of Neurobiology*, *60*, 236–248.

Cover, K. K., Maeng, L. Y., Lebrón-Milad, K., & Milad, M. R. (2014). Mechanisms of estradiol in fear circuitry: Implications for sex differences in psychopathology. *Translational Psychiatry*, *4*, e422. https://doi.org/10.1038/tp.2014.67.

Devilbiss, D. M., Jenison, R. L., & Berridge, C. W. (2012). Stress-induced impairment of a working memory task: Role of spiking rate and spiking history predicted discharge. *PLoS Computational Biology*, *8*(9), e1002681.

Fiorenza, N. G., Rosa, J., Izquierdo, I., & Myskiw, J. C. (2012). Modulation of the extinction of two different fear-motivated tasks in three distinct brain areas. *Behavioural Brain Research*, *232*(1), 210–216. https://doi.org/10.1016/j.bbr.2012.04.015.

Garcia, R., Spennato, G., Nilsson-Todd, L., Moreau, J.-L., & Deschaux, O. (2008). Hippocampal low-frequency stimulation and chronic mild stress similarly disrupt fear extinction memory in rats. *Neurobiology of Learning and Memory*, *89*(4), 560–566.

Garrett, J. E., & Wellman, C. L. (2009). Chronic stress effects on dendritic morphology in medial prefrontal cortex: Sex differences and estrogen dependence. *Neuroscience*, *162*, 195–207.

Godar, S. C., Bortolato, M., Richards, S. E., Li, F. G., Chen, K., Wellman, C. L., et al. (2015). Monoamine oxidase A is required for rapid dendritic remodeling in response to stress. *International Journal of Neuropsychopharmacology*, *18*(9), 1–12. https://doi.org.proxy.ulib.uits.iu.edu/10.1093/ijnp/pyv035.

Gordon, U., Polsky, A., & Schiller, J. (2006). Plasticity compartments in basal dendrites of neocortical pyramidal neurons. *Journal of Neuroscience*, *26*(49), 12717–12726. https://doi.org/10.1523/Jneurosci.3502-06.2006.

Goswami, S., Cascardi, M., Rodriguez-Sierra, O. E., Duvarci, S., & Pare, D. (2010). Impact of predatory threat on fear extinction in Lewis rats. *Learning & Memory*, *17*(10), 494–501. https://doi.org/10.1101/lm.1948910.

Gourley, S. L., Kedves, A. T., Olausson, P., & Taylor, J. R. (2009). A history of corticosterone exposure regulates fear extinction and cortical NR2B, GluR2/3, and BDNF. *Neuropsychopharmacology*, *34*(3), 707–716.

Gourley, S. L., Swanson, A. M., & Koleske, A. J. (2013). Corticosteroid-induced neural remodeling predicts behavioral vulnerability and resilience. *The Journal of Neuroscience*, *33*(7), 3107–3112. https://doi.org/10.1523/jneurosci.2138-12.2013.

Govindarajan, A., Rao, B. S., Nair, D., Trinh, M., Mawjee, N., Tonegawa, S., et al. (2006). Transgenic brain-derived neurotrophic factor expression causes both anxiogenic and antidepressant effects. *Proceedings of the National Academy of Science*, *103*, 13208–13213.

Graybeal, C., Feyder, M., Schulman, E., Saksida, L. M., Bussey, T. J., Brigman, J. L., et al. (2011). Paradoxical reversal learning enhancement by stress or prefrontal cortical damage: Rescue with BDNF. *Nature Neuroscience*, *14*(12), 1507–1509. https://doi.org/10.1038/nn.2954.

Groenewegen, H. J. (1988). Organization of the afferent connections of the mediodorsal thalamic nucleus in the rat, related to the mediodorsal-prefrontal topography. *Neuroscience, 24*(2), 379–431.

Hains, A. B., Vu, M. A. T., Maciejewski, P. K., van Dyck, C. H., Gottron, M., & Arnsten, A. F. T. (2009). Inhibition of protein kinase C signaling protects prefrontal cortex dendritic spines and cognition from the effects of chronic stress. *Proceedings of the National Academy of Sciences of the United States of America, 106*(42), 17957–17962.

Hamilton, D. A., & Brigman, J. L. (2015). Behavioral flexibility in rats and mice: Contributions of distinct frontocortical regions. *Genes, Brain and Behavior, 14*, 4–21.

Han, E. B., & Heinemann, S. F. (2013). Distal dendritic inputs control neuronal activity by heterosynaptic potentiation of proximal inputs. *The Journal of Neuroscience, 33*(4), 1314–1325. https://doi.org/10.1523/jneurosci.3219-12.2013.

Hikind, M., & Maroun, M. (2008). Microinfusion of the D1 receptor antagonist, SCH23390 into the IL but not the BLA impairs consolidation of extinction of auditory fear conditioning. *Neurobiology of Learning and Memory, 90*, 217–222.

Hinwood, M., Morandini, J., Day, T. A., & Walker, F. R. (2012). Evidence that microglia mediate the neurobiological effects of chronic psychological stress on the medial prefrontal cortex. *Cerebral Cortex, 22*(6), 1442–1454. https://doi.org/10.1093/cercor/bhr229.

Hoffman, A. N., Armstrong, C. E., Hanna, J. J., & Conrad, C. D. (2010). Chronic stress, cyclic 17β-estradiol, and daily handling influences on fear conditioning in the female rat. *Neurobiology of Learning and Memory, 94*, 422–433.

Holmes, A., & Wellman, C. L. (2009). Stress-induced prefrontal reorganization and executive dysfunction in rodents. *Neuroscience and Biobehavioral Reviews, 33*, 773–783.

Horchar, M. J., & Wohleb, E. S. (2019). Glucocorticoid receptor antagonism prevents microglia-mediated neuronal remodeling and behavioral despair following chronic unpredictable stress. *Brain, Behavior, and Immunity, 81*, 329–340.

Hurtubise, J. L., & Howland, J. G. (2017). Effects of stress on behavioral flexibility in rodents. *Neuroscience, 345*, 176–192. https://doi.org/10.1016/j.neuroscience.2016.04.007.

Izquierdo, A., Wellman, C. L., & Holmes, A. (2006). Brief uncontrollable stress causes dendritic retraction in infralimbic cortex and resistance to fear extinction in mice. *Journal of Neuroscience, 26*(21), 5733–5738.

Jett, J. D., Bulin, S. E., Hatherall, L. C., McCartney, C. M., & Morilak, D. A. (2017). Deficits in cognitive flexibility induced by chronic unpredictable stress are associated with impaired glutamate neurotransmission in the rat medial prefrontal cortex. *Neuroscience, 346*, 284–297. https://doi.org/10.1016/j.neuroscience.2017.01.017.

Kesner, R. P., & Churchwell, J. C. (2011). An analysis of rat prefrontal cortex in mediating executive function. *Neurobiology of Learning and Memory, 96*(3), 417–431.

Kim, D. J., St Louis, N., Molaro, R. A., Hudson, G. T., Chorley, R. C., & Anderson, B. J. (2017). Repeated unpredictable threats without harm impair spatial working memory in the Barnes maze. *Neurobiology of Learning and Memory, 137*, 92–100. https://doi.org/10.1016/j.nlm.2016.11.014.

Knapska, E., Macias, M., Mikosz, M., Nowak, A., Owczarek, D., Wawrzyniak, M., et al. (2012). Functional anatomy of neural circuits regulating fear and extinction. *Proceedings of the National Academy of Sciences of the United States of America, 109*(42), 17093–17098. https://doi.org/10.1073/pnas.1202087109.

Knapska, E., & Maren, S. (2009). Reciprocal patterns of c-Fos expression in the medial prefrontal cortex and amygdala after extinction and renewal of conditioned fear. *Learning & Memory, 16*(8), 486–493. https://doi.org/10.1101/lm.1463909.

Knox, D., George, S. A., Fitzpatrick, C. J., Rabinak, C. A., Maren, S., & Liberzon, I. (2012). Single prolonged stress disrupts retention of extinguished fear in rats. *Learning & Memory, 19*(2), 43–49. https://doi.org/10.1101/lm.024356.111.

Knox, D., Nault, T., Henderson, C., & Liberzon, I. (2012). Glucocorticoid receptors and extinction retention deficits in the single prolonged stress model. *Neuroscience, 223*, 163–173.

Knox, D., Perrine, S. A., George, S. A., Galloway, M. P., & Liberzon, I. (2010). Single prolonged stress decreases glutamate, glutamine, and creatine concentrations in the rat medial prefrontal cortex. *Neuroscience Letters, 480*(1), 16–20. https://doi.org/10.1016/j.neulet.2010.05.052.

Knox, D., Stanfield, B. R., Staib, J. M., David, N. P., DePietro, T., Chamness, M., et al. (2018). Using c-Jun to identify fear extinction learning-specific patterns of neural activity that are affected by single prolonged stress. *Behavioural Brain Research, 341*, 189–197. https://doi.org/10.1016/j.bbr.2017.12.037.

Lechin, F., Van der Dijs, B., & Benaim, M. (1996). Stress versus depression. *Progress in Neuro-Psychopharmacology & Biological Psychiatry, 20*(6), 899–950.

Lee, Y.-A., & Goto, Y. (2011). Chronic stress modulation of prefrontal cortical NMDA receptor expression disrupts limbic structure–prefrontal cortex interaction. *European Journal of Neuroscience, 34*(3), 426–436. https://doi.org/10.1111/j.1460-9568.2011.07750.x.

Li, N., Liu, R.-J., Dwyer, J. M., Banasr, M., Lee, B., Son, H., et al. (2011). Glutamate N-methyl-D-aspartate receptor antagonists rapidly reverse behavioral and synaptic deficits caused by chronic stress exposure. *Biological Psychiatry, 59*, 754–761.

Lin, G. L., Borders, C. B., Lundewall, L. J., & Wellman, C. L. (2015). D1 receptors regulate dendritic morphology in normal and stressed prelimbic cortex. *Psychoneuroendocrinology, 51*, 101–111.

Lin, Y., Ter Horst, G. J., Wichmann, R., Bakker, P., Liu, A. H., Li, X. J., et al. (2009). Sex differences in the effects of acute and chronic stress and recovery after long-term stress on stress-related brain regions of rats. *Cerebral Cortex, 19*(9), 1978–1989. https://doi.org/10.1093/cercor/bhn225.

Liston, C., McEwen, B. S., & Casey, B. J. (2009). Psychosocial stress reversibly disrupts prefrontal processing and attentional control. *Proceedings of the National Academy of Sciences of the United States of America, 106*(3), 912–917. https://doi.org/10.1073/pnas.0807041106.

Liston, C., Miller, M. M., Goldwater, D. S., Radley, J. J., Rocher, A. B., Hof, P. R., et al. (2006). Stress-induced alterations in prefrontal cortical dendritic morphology predict selective impairments in perceptual attentional set-shifting. *Journal of Neuroscience, 26*(30), 7870–7874. https://doi.org/10.1523/jneurosci.1184-06.2006.

Liu, R.-J., & Aghajanian, G. K. (2008). Stress blunts serotonin- and hypocretin-evoked EPSCs in prefrontal cortex: Role of corticosterone-mediated apical dendritic atrophy. *Proceedings of the National Academy of Sciences, 105*(1), 359–364. https://doi.org/10.1073/pnas.0706679105.

Liu, R.-J., Lee, F. S., Li, X.-Y., Bambico, F., Duman, R. S., & Aghajanian, G. K. (2012). Brain-derived neurotrophic factor Val66Met allele impairs basal and ketamine-stimulated synaptogenesis in prefrontal cortex. *Biological Psychiatry, 71*(11), 996–1005.

Maren, S., & Holmes, A. (2016). Stress and fear extinction. *Neuropsychopharmacology Reviews, 41*, 58–79. https://doi.org/10.1038/npp.2015.180.

Maroun, M., Ioannides, P. J., Bergman, K. L., Kavushansky, A., Holmes, A., & Wellman, C. L. (2013). Fear extinction deficits following acute stress associate with increased spine density and dendritic retraction in basolateral amygdala neurons. *European Journal of Neuroscience, 38*, 2611–2620.

Maroun, M., & Richter-Levin, G. (2003). Exposure to acute stress blocks the induction of long-term potentiation of the amygdala-prefrontal cortex pathway in vivo. *Journal of Neuroscience, 23*(11), 4406–4409.

Martin, K. P., & Wellman, C. L. (2011). NMDA receptor blockade alters stress-induced dendritic remodeling in medial prefrontal cortex. *Cerebral Cortex, 21*(10), 2366–2373. https://doi.org/10.1093/cercor/bhr021.

McEwen, B. S. (2017). Allostasis and the epigenetics of brain and body health over the life course: The brain on stress. *JAMA Psychiatry*, *74*(6), 551–552. https://doi.org/10.1001/jamapsychiatry.2017.0270.

McGinty, J. F., Whitfield, T. W., & Berglind, W. J. (2010). Brain-derived neurotrophic factor and cocaine addiction. *Brain Research*, *1314*, 183–193. https://doi.org/10.1016/j.brainres.2009.08.078.

McKlveen, J. M., Morano, R. L., Fitzgerald, M., Zoubovsky, S., Cassella, S. N., Scheimann, J. R., et al. (2016). Chronic stress increases prefrontal inhibition: A mechanism for stress-induced prefrontal dysfunction. *Biological Psychiatry*, *80*(10), 754–764. https://doi.org/10.1016/j.biopsych.2016.03.2101.

McKlveen, J. M., Myers, B., & Herman, J. P. (2015). The medial prefrontal cortex: Coordinator of autonomic, neuroendocrine and behavioural responses to stress. *Journal of Neuroendocrinology*, *27*(6), 446–456. https://doi.org/10.1111/jne.12272.

Meaney, M. J., & Aitken, D. H. (1985). [3H]Dexamethasone binding in rat frontal cortex. *Brain Research*, *328*(1), 176–180 (doi:0006-8993(85)91340-X [pii).

Mika, A., Mazur, G. J., Hoffman, A. N., Talboom, J. S., Bimonte-Nelson, H. A., Sanabria, F., et al. (2012). Chronic stress impairs prefrontal cortex-dependent response inhibition and spatial working memory. *Behavioral Neuroscience*, *126*(5), 605–619. https://doi.org/10.1037/a0029642.

Miracle, A. D., Brace, M. F., Huyck, K. D., Singler, S. A., & Wellman, C. L. (2006). Chronic stress impairs recall of extinction of conditioned fear. *Neurobiology of Learning and Memory*, *85*, 213–218.

Mizoguchi, K., Yuzurihara, M., Ishige, A., Sasaki, H., Chui, D.-H., & Tabira, T. (2000). Chronic stress induces impairment of spatial working memory because of prefrontal dopaminergic dysfunction. *Journal of Neuroscience*, *20*(4), 1568–1574.

Moench, K. M., Breach, M. R., & Wellman, C. L. (2019). Chronic stress produces enduring sex- and region-specific alterations in novel stress-induced c-Fos expression. *Neurobiology of Stress*, *10*, 100147. https://doi.org/10.1016/j.ynstr.2019.10014.

Moench, K. M., Breach, M. R., & Wellman, C. L. (2020). Prior stress followed by a novel stress challenge results in sex-specific deficits in behavioral flexibility and changes in gene expression in rat medial prefrontal cortex. *Hormones and Behavior*, *117*, 104615. https://doi.org/10.1016/j.yhbeh.2019.104615.

Moench, K. M., Maroun, M., Kavushansky, A., & Wellman, C. L. (2016). Alterations in neuronal morphology in infralimbic cortex predict resistance to fear extinction following acute stress. *Neurobiology of Stress*, *3*, 23–33. https://doi.org/10.1016/j.ynstr.2015.12.002.

Moench, K. M., & Wellman, C. L. (2017). Differential dendritic remodeling in prelimbic cortex of male and female rats during recovery from chronic stress. *Neuroscience*, *357*, 145–159.

Nikiforuk, A. (2012). Dopamine D1 receptor modulation of set shifting: The role of stress exposure. *Behavioural Pharmacology*, *23*(4), 434–438. https://doi.org/10.1097/FBP.0b013e328356522f.

Nikiforuk, A., & Popik, P. (2011). Long-lasting cognitive deficit induced by stress is alleviated by acute administration of antidepressants. *Psychoneuroendocrinology*, *36*(1), 28–39. https://doi.org/10.1016/j.psyneuen.2010.06.001.

Nikiforuk, A., & Popik, P. (2013). Neurochemical modulation of stress-induced cognitive inflexibility in a rat model of an attentional set-shifting task. *Pharmacological Reports*, *65*(6), 1479–1488.

Ostrander, M. M., Ulrich-Lai, Y. M., Choi, D. C., Flak, J. N., Richtand, N. M., & Herman, J. P. (2009). Chronic stress produces enduring decreases in novel stress-evoked c-fos mRNA expression in discrete brain regions of the rat. *Stress—The International Journal on the Biology of Stress*, *12*(6), 469–477. https://doi.org/10.3109/10253890802641966.

Otis, J. M., Fitzgerald, M. K., & Mueller, D. (2014). Infralimbic BDNF/TrkB enhancement of GluN2B currents facilitates extinction of a cocaine-conditioned place preference. *Journal of Neuroscience*, *34*(17), 6057–6064. https://doi.org/10.1523/Jneurosci.4980-13.2014.

Papaleo, F., Silverman, J. L., Aney, J., Tian, Q., Barkan, C. L., Chadman, K. K., et al. (2011). Working memory deficits, increased anxiety-like traits, and seizure susceptibility in BDNF overexpressing mice. *Learning & Memory*, *18*(8), 534–544. https://doi.org/10.1101/lm.2213711.

Peters, J., Dieppa-Perea, L. M., Melendez, L. M., & Quirk, G. J. (2010). Induction of fear extinction with hippocampal-infralimbic BDNF. *Science*, *328*(5983), 1288–1290. https://doi.org/10.1126/science.1186909.

Popoli, M., Yan, Z., McEwen, B. S., & Sanacora, G. (2012). The stressed synapse: The impact of stress and glucocorticoids on glutamate transmission. *Nature Reviews Neuroscience*, *13*(1), 22–37. http://www.nature.com/nrn/journal/v13/n1/suppinfo/nrn3138_S1.html.

Radley, J. J., Anderson, R. M., Hamilton, B. A., Alcock, J. A., & Romig-Martin, S. A. (2013). Chronic stress-induced alterations of dendritic spine subtypes predict functional decrements in an hypothalamo-pituitary-adrenal-inhibitory prefrontal circuit. *The Journal of Neuroscience: The Official Journal of the Society for Neuroscience*, *33*(36), 14379–14391.

Radley, J. J., Rocher, A. B., Janssen, W. G. M., Hof, P. R., McEwen, B. S., & Morrison, J. H. (2005). Reversibility of apical dendritic retraction in the rat medial prefrontal cortex following repeated stress. *Experimental Neurology*, *196*(1), 199–203.

Radley, J. J., Rocher, A. B., Miller, M., Janssen, W. G. M., Liston, C., Hof, P. R., et al. (2006). Repeated stress induces dendritic spine loss in the rat medial prefrontal cortex. *Cerebral Cortex*, *16*(3), 313–320.

Radley, J. J., Rocher, A. B., Rodriguez, A., Ehlenberger, D. B., Dammann, M., McEwen, B. S., et al. (2008). Repeated stress alters dendritic spine morphology in the rat medial prefrontal cortex. *The Journal of Comparative Neurology*, *507*(1), 1141–1150.

Radley, J. J., Sisti, H. M., Hao, J., Rocher, A. B., McCall, T., Hof, P. R., et al. (2004). Chronic behavioral stress induces apical dendritic reorganization in pyramidal neurons of the medial prefrontal cortex. *Neuroscience*, *125*(1), 1–6.

Rall, W., Burke, R. E., Holmes, W. R., Jack, J. J., Redman, S. J., & Segev, I. (1992). Matching dendritic neuron models of experimental data. *Physiological Reviews*, *72*, S159–S186.

Risch, N., Herrell, R., Lehner, T., Liang, K. Y., Eaves, L., Hoh, J., et al. (2009). Interaction between the serotonin transporter gene (5-HTTLPR), stressful life events, and risk of depression a meta-analysis. *JAMA—Journal of the American Medical Association*, *301*(23), 2462–2471.

Rosas-Vidal, L. E., Do-Monte, F. H., Sotres-Bayon, F., & Quirk, G. J. (2014). Hippocampal-prefrontal BDNF and memory for fear extinction. *Neuropsychopharmacology*, *39*(9), 2161–2169. https://doi.org/10.1038/npp.2014.64.

Salter, M. W., & Beggs, S. (2014). Sublime microglia: Expanding roles for the guardians of the CNS. *Cell*, *58*, 15–24.

Salter, M. W., & Stevens, B. (2017). Microglia emerge as central players in brain disease. *Nature Medicine*, *23*, 1018. https://doi.org/10.1038/nm.4397.

Santini, E., Muller, R. U., & Quirk, G. J. (2001). Consolidation of extinction learning involves transfer from NMDA-independent to NMDA-dependent memory. *Journal of Neuroscience*, *21*(22), 9009–9017.

Sapolsky, R. M. (2003). Stress and plasticity in the limbic system. *Neurochemical Research*, *28*(11), 1735–1742.

Segal, M. (2017). Dendritic spines: Morphological building blocks of memory. *Neurobiology of Learning and Memory*, *138*, 3–9. https://doi.org/10.1016/j.nlm.2016.06.007.

Seib, L. M., & Wellman, C. L. (2003). Daily injections alter spine density in rat medial prefrontal cortex. *Neuroscience Letters*, *337*, 29–32.

Shansky, R. M., Hamo, C., Hof, P. R., Lou, W., McEwen, B. S., & Morrison, J. H. (2010). Estrogen promotes stress sensitivity in a prefrontal cortex–amygdala pathway. *Cerebral Cortex*, *20*(11), 2560–2567. https://doi.org/10.1093/cercor/bhq003.

Shansky, R. M., Hamo, C., Hof, P. R., McEwen, B. S., & Morrison, J. H. (2009). Stress-induced dendritic remodeling in the prefrontal cortex is circuit specific. *Cerebral Cortex*, *19*(10), 2479–2484. https://doi.org/10.1093/cercor/bhp003.

Shepard, R., & Coutellier, L. (2018). Changes in the prefrontal glutamatergic and parvalbumin systems of mice exposed to unpredictable chronic stress. *Molecular Neurobiology*, *55*(3), 2591–2602.

Shepard, R., Page, C. E., & Coutellier, L. (2016). Sensitivity of the prefrontal GABAergic system to chronic stress in male and female mice: Relevance for sex differences in stress-related disorders. *Neuroscience*, *332*, 1–12.

Sinclair, D., Purves-Tyson, T., Allen, K., & Weickert, C. (2014). Impacts of stress and sex hormones on dopamine neurotransmission in the adolescent brain. *Psychopharmacology*, *231*(8), 1581–1599. https://doi.org/10.1007/s00213-013-3415-z.

Spruston, N. (2008). Pyramidal neurons: Dendritic structure and synaptic integration. *Nature Reviews Neuroscience*, *9*(3), 206–221. https://doi.org/10.1038/nrn2286.

Swanson, L. W., & Cowan, W. M. (1977). An autoradiographic study of the organization of the efferent connections of the hippocampal formation in the rat. *Journal of Comparative Neurology*, *172*(1), 49–84.

Swanson, A. M., Shapiro, L. P., Whyte, A. J., & Gourley, S. L. (2013). Glucocorticoid receptor regulation of action selection and prefrontal cortical dendritic spines. *Communicative & Integrative Biology*, *6*(6), e26068. https://doi.org/10.4161/cib.26068.

Tritsch, N. X., & Sabatini, B. L. (2012). Dopaminergic modulation of synaptic transmission in cortex and striatum. *Neuron*, *76*(1), 33–50. https://doi.org/10.1016/j.neuron.2012.09.023.

Turner, R. J., & Lloyd, D. A. (2004). Stress burden and the lifetime incidence of psychiatric disorder in young adults: Racial and ethnic contrasts. *Archives of General Psychiatry*, *61*(5), 481–488. https://doi.org/10.1001/archpsyc.61.5.481.

Urban-Ciecko, J., & Barth, A. L. (2016). Somatostatin-expressing neurons in cortical networks. *Nature Reviews Neuroscience*, *17*(7), 401–409. https://doi.org/10.1038/nrn.2016.53.

Vieira, P. A., Corches, A., Lovelace, J. W., Westbrook, K. B., Mendoza, M., & Korzus, E. (2015). Prefrontal NMDA receptors expressed in excitatory neurons control fear discrimination and fear extinction. *Neurobiology of Learning and Memory*, *119*, 52–62. https://doi.org/10.1016/j.nlm.2014.12.012.

Vijayraghavan, S., Wang, M., Birnbaum, S. G., Williams, G. V., & Arnsten, A. F. T. (2007). Inverted-U dopamine D1 receptor actions on prefrontal neurons engaged in working memory. *Nature Neuroscience*, *10*(3), 376–384. http://www.nature.com/neuro/journal/v10/n3/suppinfo/nn1846_S1.html.

Walker, F. R., Nilsson, M., & Jones, K. (2013). Acute and chronic stress-induced disturbances of microglial plasticity, phenotype and function. *Current Drug Targets*, *14*(11), 1262–1276.

Wei, J., Yuen, E. Y., Liu, W., Li, X., Zhong, P., Karatsoreos, I. N., et al. (2014). Estrogen protects against the detrimental effects of repeated stress on glutamatergic transmission and cognition. *Molecular Psychiatry*, *19*(5), 588–598. https://doi.org/10.1038/mp.2013.83.

Wellman, C. L. (2001). Dendritic reorganization in pyramidal neurons in medial prefrontal cortex after chronic corticosterone administration. *Journal of Neurobiology*, *49*(3), 245–253.

Wellman, C. L., & Moench, K. M. (2019). Preclinical studies of stress, extinction, and prefrontal cortex: Intriguing leads and pressing questions. *Psychopharmacology*, *236*, 59–72.

Wilber, A. A., Walker, A. G., Southwood, C. J., Farrell, M. R., Lin, G. L., Rebec, G. V., et al. (2011). Chronic stress alters neural activity in medial prefrontal cortex during retrieval of extinction. *Neuroscience*, *174*, 115–131.

Wohleb, E. S., Terwilliger, R., Duman, C. H., & Duman, R. S. (2018). Stress-induced neuronal CSF1 provokes microglia-mediated neuronal remodeling and depressive-like behavior. *Biological Psychiatry*, *83*, 38–49. https://doi.org/10.1016/j.biopsych.2017.05.026.

Yu, H., Wang, D.-D., Wang, Y., Liu, T., Lee, F. S., & Chen, Z.-Y. (2012). Variant brain-derived neurotrophic factor Val66Met polymorphism alters vulnerability to stress and response to antidepressants. *The Journal of Neuroscience: The Official Journal of the Society for Neuroscience*, *32*(12), 4092–4101.

Yuen, E. Y., Liu, W., Karatsoreos, I. N., Feng, J., McEwen, B. S., & Yan, Z. (2009). Acute stress enhances glutamatergic transmission in prefrontal cortex and facilitates working memory. *Proceedings of the National Academy of Sciences of the United States of America*, *106*(33), 14075–14079.

Yuen, E. Y., Wei, J., Liu, W., Zhong, P., Li, X., & Yan, Z. (2012). Repeated stress causes cognitive impairment by suppressing glutamate receptor expression and function in prefrontal cortex. *Neuron*, *73*, 962–977.

CHAPTER SEVEN

Environment and early life: Decisive factors for stress-resilience and vulnerability

Akshaya Hegde, Rupshi Mitra*
School of Biological Sciences, Nanyang Technological University, Singapore, Singapore
*Corresponding author: e-mail address: rmitra@ntu.edu.sg

Contents

Abstract

Early life is a critical and sensitive period whereby environmental imprints on later life are generated. These environmental influences from early life have long-lasting consequences on for mental health. Both human and animal work suggests that maternal presence constitutes an important part of the early experience. Maternal separation causes a sustained increase in stress responsiveness later in life, along with facilitated anxiety-like behaviors. On the contrary, providing a complex and enriching sensory environment during or after stress, on the other hand, creates resilience to stress. In this chapter, we summarize these environmental influences on the maternal interactions and subsequent stress susceptibility or resilience of the offspring.

International Review of Neurobiology, Volume 150
ISSN 0074-7742
https://doi.org/10.1016/bs.irn.2019.12.002

1. The influence of early life stress in behavioral plasticity

Individual animals show remarkable variance in their behavior, which can be largely influenced by the incipient environment during the early period of their life. Behavioral and physiological phenotypes for the same genetic backgrounds are often changeable or plastic in response to changes in the environment (Gilbert, 2001, 2004; Sultan, 2003). Pigliucci, Murren, and Schlichting (2006) defined *phenotypic plasticity* as "the ability of a single genotype to give rise to various phenotypes when exposed to differential environmental conditions" (Pigliucci et al., 2006). This involves the possibility to change the developmental pathway (*developmental plasticity*) of an individual in response to a particular environment cue and also the ability of an individual to alter their phenotypic activity (e.g., metabolism) in response to different environmental conditions (Garland & Kelly, 2006). In general, the earlier an organism's development is disturbed, higher is the amplitude of detrimental effects (Desai & Hales, 1997; Henry & Ulijaszek, 1996). With the advent of increasing urbanization and expanding pace of technology-driven change in the living environment, studying the relationship between the environment and the phenotypic plasticity is of enormous importance. Animals have evolved by adapting themselves to various dynamic environments through temporary or permanent changes in the physiology and behavior of an individual. The early environment plays a dominant part in this environment-dependent plasticity. For example, variation in maternal care received during early life creates corresponding variation in maternal care exhibited as an adult in rodents (Cameron et al., 2008). This parental entrainment of nursing allows rodents to shift between a high-care and low-care styles depending on the extant environment. Similarly, stress hormones produced by squirrel mothers in response to high population density increase growth rate of offspring (Dantzer et al., 2013), resulting in a life history more suited to intra-species competition.

During early life, the time window that is vulnerable to the changing environment is commonly referred to as the "*critical*" or "*sensitive*" period (Tau & Peterson, 2010). An organism requires various stimuli of visual, olfactory, and social cues from the surroundings for its normative development during the sensitive period (Knudsen, 2004). Observations show that during the sensitive period, development of an individual depends on the type of cue perceived from the environment (Tierney & Nelson III, 2009).

A developmental trajectory established during a sensitive period in early life is typically challenging to reverse in the later period (Bateson, 2001). If the phenotype obtained in early life is not compatible with the conditions later in life, developmental plasticity might have detrimental effects on the survival and future reproductive performance of the individual (McNamara & Houston, 1996; Moran, 1992; Sultan & Spencer, 2002; Tufto, 2000). Thus, it is apparent that the environment during the *sensitive period* plays a crucial role in the healthy development of an individual.

Among the numerous factors available in the environment, one of fundamental essence during an individual's early life is "mother." In several bird species, maternal influences can change multiple components of eggs, e.g., size, hormonal content, nutritional availability, and sex of the embryo (Groothuis, Müller, von Engelhardt, Carere, & Eising, 2005; Mazuc, Chastel, & Sorci, 2003; Rutstein, Gilbert, Slater, & Graves, 2004). The availability of food to the mothers or quality of their mating partners can cause variations in maternal influences in many birds and mammals. This variation has a significant impact on the developmental phenotypic plasticity of their offspring that persist across various generations, even if the conditions have changed later on (Lindström, 1999; Mousseau & Fox, 1998). Mothers in numerous vertebrates have an enormous impact on growth and breeding performance of their offspring, depending on their own early life experience (Kan, Callaghan, & Richardson, 2016). For example, mothers in mammals that have undergone food restriction in their early life give birth to offspring that have low birth-weight (William Huck, Labov, & Lisk, 1987). This shows that the interaction between a mother and the offspring plays a vital role in its development process.

2. A critical role of mother-infant interactions in mental health

In embryonic life, the prenatal environment within the mother's womb also plays a crucial role in an offspring's survival and reproductive success. In humans, studies have shown that the brain morphology and physiology of the embryo depends on the nutritional circumstances of the mother (Barker, 1998; Barker, Eriksson, Forsén, & Osmond, 2002). This, in turn, prepares the baby for an environment that it is likely to be born into. If the "prenatal conditions" and "early-life conditions" of a baby are in contrast with each other, the consequences could be harmful (Bateson, 2001). It has been observed that babies with low birth weight due to stressful prenatal periods that grew up in a favorable environment, later develop multiple health problems, including heart disease, diabetes, and hypertension (Barker, 1998; Gluckman & Hanson, 2004; Godfrey, Barker, Robinson, & Osmond, 1997). However, if the birth weight is high and the later life has a favorable environment, the risk of having health problems remains low (Kelly, Nazroo, McMunn, Boreham, & Marmot, 2001). Therefore, the cues an individual receives prenatally has a significant influence on the early development and health success (Eriksson et al., 1999). A classic illustrator of this idea is the Dutch hunger during the Second World War. Pregnant women in the Dutch famine in 1944 underwent starvation, and this caused an adverse fetal environment due to lack of nutrition and energy supplements. These harsh prenatal conditions caused low bodyweight of the babies at birth, and these individuals were susceptible to compromised cognitive functions, cardiovascular disease, and obesity despite adequate food intake later in life. This was due to the food-deprived environment that persisted during their fetal stage, which programmed their bodies to conserve nutrients throughout the course of their life (Schulz, 2010). The information above shows that early life focuses on preparing an offspring for a future environment based on the cues it receives during the crucial period of development even before birth.

The mother-infant relationship during postnatal life also plays a cardinal role in phenotypic plasticity. Bowlby stated that a mother's love during early life is as crucial for "mental health" as proteins and vitamins are for "physical health" (Bowlby, 1958; Rutter, 1972). Multiple studies have focused on the importance of the mother-infant relationship for the behavioral and cognitive development of infants (Ainsworth & Bell, 1970; Bowlby, 1958; Maccoby & Masters, 1970). Studies have shown that postpartum depression in mothers can compromise cognitive and emotional development of the child, perhaps due to disruption of positive social interactions (Righetti-Veltema, Conne-Perréard, Bousquet, & Manzano, 2002; Shaffer & Kipp, 2002; Sroufe, 1997). Similarly, early-life stress of mothers due to childhood abuse results in later adverse effects on the development of their child (Murray, Stanley, Hooper, King, & Fiori-Cowley, 1996; Muzik et al., 2013). Therefore, mother-infant relationships have a long-term effect on future health of a child. A positive mother-infant relationship during the postnatal environment is beneficial for development of the affective behavior of the infant (Baydar & Brooks-Gunn, 1991; Field, Healy, Goldstein, & Guthertz, 1990). Therefore, the manner in which an infant establishes the relationship with the mother influences the manner in which it establishes relationships with every other facet in the environment throughout the rest of her/his life (Ainsworth, 1979). Due to the importance of the mother-child relationship to human health, this phenomenon has been extensively modeled in rodents where experimental investigations are possible.

Work by Michael Meaney and colleagues confirms that alteration in maternal care can have a significant impact on the physiological development of the offspring. The neuroendocrine axis comprising hypothalamus, pituitary, and adrenal (HPA-axis) coordinates physiological changes in rodents (Fig. 1). The HPA axis releases glucocorticoids that mediate the stress response, helping body to return to physiological balance after an encounter of stress episodes. Rat offspring that receive "high maternal care" in the form of licking and grooming from the mother during initial 10 postnatal days display low glucocorticoid response to acute stress during adulthood. These adults also express more glucocorticoid receptors messenger RNA (mRNA) when compared to their counterparts with "low maternal care," thus providing a more potent negative feedback on the further activity of HPA-axis. Thus, the amount of maternal care in rats during that early stage is crucial in programming the adult stress physiology of the individual (Liu et al., 1997). Adverse conditions in early life can cause sustained behavioral deficits and molecular changes within brain that persist through later life (Suderman et al., 2012). Mere maternal presence can alter the behavioral development of progeny. Regina Sullivan and colleagues, for example, show rat pups in the presence of mothers exhibit an atypical preference to odors that have been previously conditioned using paired odor-footshock training (Moriceau & Sullivan, 2006). This atypical preference to a fear-conditioned odor requires maternal presence, as shown by resumption of aversion in when mother is absent during the training. Maternal presence, in this case, shifts offspring behavior through preventing interaction of stress hormones and developing amygdala.

The examples in this section demonstrate that disruption of maternal interactions can predispose offspring to deleterious effects. Moreover, it shows that the maintenance of healthy mother-child relationships within society might be important to create a healthy and resilient society. Secure attachment in early life can lead to better emotional outcomes (Bohlin, Hagekull, & Rydell, 2000) and better cognition later in life (Schore, 2001). Similarly, effects of genes that create a predisposition to an overactive stress response can be moderated by maternal buffering (Suomi, 2011). Thus, healthy interactions in early life establish a resilient individual phenotype.

3. Mechanisms of homeostasis, allostatic load, stress vulnerability, and resilience

Homeostasis originated from the Greek words "same" and "steady," which means maintaining a steady and stable state of vital physiological processes that are important for an individual's survival. Walter Cannon coined this term in 1930 and described it as various ways a body maintains steady states of various vital functions like pH, temperature, or glucose levels in response to varying environments (Walter Bradford Cannon, 1932a, 1932b). Allostasis is a related, but distinct, component from homeostasis. Allostasis refers to the active processes that the body undergoes to reestablish homeostasis when exposed to a stressor. This could involve production of hormones or various mediators (e.g., adrenaline, noradrenaline, glucocorticoids, cytokine) that help an organism to adapt to different situations (McEwen, 1998a, 2003; McEwen & Wingfield, 2003). If heart rate is considered as an example, homeostasis refers to maintaining the heart rate at a stable "set point" for survival, and allostasis is elevating the heart rate with the help of mediators (adrenaline) to adapt to a challenging situation (stress). The allostatic state promotes adaptation for the short-term, but when this occurs chronically, it leads to a cost or an allostatic load that generates disease like hypertension due to chronic levels of elevated heart rate and constricted blood vessels. Therefore, chronic stress leads to detrimental problems in an individual due to allostatic load (Korte, Koolhaas, Wingfield, & McEwen, 2005; McEwen & Wingfield, 2003).

Stress is defined as a condition that disturbs the physiological or psychological homeostasis of an individual. Walter Bradford Cannon (1932a, 1932b) and Selye (1936) proposed an early conception of stress and its consequence on humans. In their theories, stress was described as a behavioral and physiological response due to interaction with the environment for the first time in the 1930s.

The excessive degree of stress leads to adverse social, psychological, and biological outcomes. For example, stress is a major predisposing factor for anxiety disorders, post-traumatic stress disorder (PTSD), and major depressive disorders (MDD) (Cohen, Janicki-Deverts, & Miller, 2007; Nestler et al., 2002). Chronic exposure to stressful events causes surge in secretion of stress hormones (Glucocorticoids, Fig. 1), leading to negative impact on brain, cognition, and behavior (Anisman, Zaharia, Meaney, & Merali, 1998; Issa, Rowe, Gauthier, & Meaney, 1990; Mitra, Jadhav, McEwen, Vyas, & Chattarji, 2005; Romeo et al., 2006;

Vallee et al., 1997). The magnitude of the effect of stress in an individual varies with the age and extent of the stress exposure. Glucocorticoids (GC) are one of the primary mediators of allostasis and are crucial to initiating stress response to cope up with a stressful situation.

Though adverse stress leads to different disorders, there are individuals who can cope better while facing negative consequences. The variation in the coping strategy that helps the organisms deal well with the environment is called "resilience." In this context, resilience refers to the capacity of an individual to successfully counter the harmful effects of stress.

Resilience is the ability of an individual to sustain homeostasis when exposed to an extraordinary level of stress and trauma (Russo, Murrough, Han, Charney, & Nestler, 2012). For example, inescapable foot shocks create learned helplessness in a majority of exposed mice, but a subset ($\approx$30%) of mice with the same treatment show active coping and attempts to escape at a level similar to non-stressed counterparts (Berton et al., 2007). Similarly, in humans, premature children born in poverty results in a majority of individuals with adverse health consequences and a minority that display resilience in terms of cognition, growth, and health (Bradley et al., 1994). Adult consequences of early life malnutrition are similarly variable between individuals (Cicchetti & Rogosch, 2009). A small fraction of individuals shows resilience without interventions. This makes it interesting to explore various ways and methods to induce resilience in vulnerable individuals for better stress coping. In other words, if we can determine why a small fraction of individuals show stress resilience in the face of adversity, we can perhaps utilize that knowledge to also promote resilience in the susceptible individuals.

As discussed earlier, both stress-vulnerability and stress-resilience are two different outcomes that can emerge from the same stressful experience. Resilience is obtained in a *small percentage* naturally in response to a stressful situation. Hence, it is necessary and vital to study the underlying neurobiological mechanisms of resilience to counter the devastating effects of stress. To determine the methods to counter stress, it is essential to understand how stress affects the biology of an individual. We will now look into the various parameters that play a crucial role in response to stress.

4. Neuroendocrine systems regulating the stress response

Activation of the Hypothalamic-Pituitary-Adrenal (HPA, Fig. 1) axis in response to stress is initiated by corticotropin-releasing factor (CRF), which is synthesized and secreted by hypophysiotropic neurons localized in the paraventricular nucleus of the hypothalamus. CRF, the principal regulator of the axis, activates CRF receptors in the anterior pituitary, causing the release of Adrenocorticotropic (ACTH) hormone into the systemic circulation. ACTH then activates the receptor on the adrenal cortex to stimulate the Glucocorticoid (GC) synthesis and further causes the release of Glucocorticoids from zona fasciculata into the bloodstream, which regulates the physiological and behavioral changes in response to stress and brings it back to baseline, when the stressor is gone. Incomplete or exaggerated activation of the HPA axis leads to emergence of pathological conditions (Bamberger, Schulte, & Chrousos, 1996; Munck, Guyre, & Holbrook, 1984). Glucocorticoids, as primary stress hormones, induce a flight or fright response and helps the body to return to homeostatic equilibrium after the stress episode is over (McEwen, 1998b).

In addition to the endocrine HPA axis, as mentioned above, there are three crucial brain regions that provide regulation of the stress response. These constituents of central "Brain-HPA" are Hippocampus, Prefrontal cortex, and Amygdala. These brain structures have a significant effect on the release of GC and behavioral response to a stressful event (McEwen, 2000; Raison & Miller, 2003).

Hippocampus is located in the medial temporal lobe of the brain and is mainly responsible for spatial, episodic, and contextual memory formation (McEwen, 2007). As shown in the figure, the hippocampus is the primary brain region that provides negative feedback to the HPA axis. Hippocampectomy, for example, results in an increased level of CRF mRNA (Herman et al., 1989; Sapolsky, Krey, & McEwen, 1984). Hippocampus is one of the most extensively explored brain regions in most of the early brain-related studies due to its extensive involvement in cognitive processes (Kim & Diamond, 2002; Lupien, McEwen, Gunnar, & Heim, 2009). Glucocorticoid (GC) cascade hypothesis proposes that an increase in glucocorticoids causes hippocampal neuronal loss, which in turn causes a reduction in negative feedback on the stress endocrine axis (Sapolsky, Krey, & McEwen, 1986). The structural plasticity of this region is highly dependent on the number and impact of stressful episodes experienced by an individual and level of glucocorticoids in the systemic circulation (McEwen, 1999, 2008). Additionally, stress-induced impairment of learning and memory is closely associated with alteration in hippocampal neurons (Magarin & McEwen, 1995; Magariños, McEwen, Flügge, & Fuchs, 1996). Early life stress induces alterations in CA3 region of hippocampal neurons (Krugers et al., 2012; Monroy, Hernández-Torres, & Flores, 2010).

Stress is known to induce dendritic hypertrophy in the basolateral amygdala (BLA) (Mitra et al., 2005; Vyas, Mitra, Rao, & Chattarji, 2002) and to enhance anxiety (Vyas et al., 2002; Vyas & Chattarji, 2004). Stress also causes an increase in the secretion of glucocorticoids (GC) that has harmful impact on an individual (Herman, Prewitt, & Cullinan, 1996; Smith & Vale, 2006). It was observed that a single acute dose of corticosterone was sufficient to produce dendritic hypertrophy in the BLA and anxiogenic behavioral output. Additionally, this pattern of dendritic hypertrophy after acute treatment was comparable to that caused by chronic treatment of corticosterone (Mitra & Sapolsky, 2008). These studies affirm the crucial role of amygdala in modulating stress response making it an exciting region to explore in early life stress models.

Unlike the hippocampus, comparatively less is known about how early life stress affects the amygdala and the nature of its role in modulating the stress response. The amygdala constitutes a group of nuclei that helps in the modulation of various functions related to initiation, processing, and consolidation of fear memory (Davis, 1992). Each nucleus of amygdala comprises different neuronal populations ranging in structure and function. The function of amygdala has been extensively studied in context of Pavlovian fear conditioning (LeDoux, 2000, 2007). The fear stimulus reaches amygdala from the thalamus or the cortex. This information is further transmitted from the basolateral amygdala (BLA) to the central amygdala that is crucial for expression of fear induced by the stimuli (LeDoux, 2003; Lin et al., 2001; Romanski & LeDoux, 1993). BLA being the primary nucleus to receive signal from the fear stimulus can be considered as an essential brain region for formation of fear-induced behavioral response. The BLA comprises seven specific types of cell populations (Faulkner & Brown, 1999), out of which the predominating pyramidal neurons, stellate neurons, and GABAergic local interneurons are three significant kinds. Pyramidal and stellate neurons are glutamatergic and excitatory in nature. GABAergic local interneurons are inhibitory in nature (Rainnie, Asprodini, & Shinnick-Gallagher, 1991a, 1991b). Activation of BLA mediates the activation of the central amygdala (CeA) of the medial amygdala (MeA) via the excitatory neurons mentioned earlier (Davis, 1998). This further causes activation of the paraventricular nucleus (PVN) of the hypothalamus via the bed nucleus of stria terminalis (BNST), resulting in facilitation of HPA axis for generating a stress response (Davis, 1992, 1998). Additionally, studying the BLA in context of early life and stress is more important than studying other cortical regions because the development of amygdala takes place during the critical early life period, which is much earlier than cortical regions (Elias & Arnold, 2006).

As mentioned earlier, BLA takes part in the consolidation of fear-induced memories by processing information from the environment (McGaugh, 2004). This highlights an interesting concept of "behavioral engagement" that sheds light on crucial processes that regulate interactions between individuals and the environment. The decision of an animal to engage or disengage from the threatening environment is extremely crucial for an animal to decide on how to respond to various situations like foraging, an unexpected encounter of predator, and other similar interactions with the environment meaningfully (Pearson, Watson, & Platt, 2014). Considering that amygdala's structural plasticity occurs much earlier in life, the engaging and disengaging behavior might also be formed, processed and organized earlier in life (Pearson et al., 2014). This makes it important to study impacts of varying environments during early life in context of the amygdala.

Bilateral radiofrequency lesions in the amygdala cause impairment in ACTH secretion (Allen & Allen, 1974, 1975). This suggests the involvement of amygdala in providing positive feedback to the HPA axis with respect to the amount of GC present in the circulation. There are other studies that support this idea wherein lesions of the central amygdala inhibits the response of the HPA axis to stressful stimuli. Additionally, the electrical stimulation of the amygdala impairs the glucocorticoid negative feedback following stressful neural stimuli (Weidenfeld & Ovadia, 2017). Amygdala is the emotional center of the brain, and it helps in formation, processing, and consolidation of fear memory (Roozendaal, McEwen, & Chattarji, 2009). Amygdala mediates fear and anxiety, and this, in turn, drives the HPA axis to release stress hormones (Davis, 1992; Davis, Rainnie, & Cassell, 1994; LeDoux, 2003; Rodrigues, LeDoux, & Sapolsky, 2009) Fear and emotion-related memories are known to be consolidated in this region which in turn helps in long-term adaptability (Rogan, Stäubli, & LeDoux, 1998). The amygdala is activated by corticosterone released during a stress response, and this causes dendritic morphological changes, and these changes have been correlated with anxiety and stress in prior studies (Mitra, Adamec, & Sapolsky, 2009; Mitra et al., 2005; Mitra & Sapolsky, 2008; Vyas et al., 2002). Neuronal morphology is associated with anxiety-like behavior and general stress response (Mitra et al., 2005; Mitra & Sapolsky, 2008; Vyas & Chattarji, 2004). Acute and/or chronic stress, as well as exogenous supplementation corticosterone, results in neuronal hypertrophy in basolateral nucleus of amygdala (BLA) (Ashokan, Hegde, & Mitra, 2016). Additionally, rats subjected to chronic immobilization stress in both early life and adulthood alter the spine population in BLA (Ashokan et al., 2016; Mitra et al., 2005).

Apart from the hippocampus and amygdala, another brain region that plays a critical role in the regulation of HPA axis is medial prefrontal cortex (mPFC). It has an inhibitory role on the axis, and the involvement of this region depends upon the magnitude of stressor. The evidence for involvement of mPFC in regulation of HPA axis was shown in studies based on lesions in mPFC that regulate ACTH and corticosterone in response to stress (Figueiredo, Bruestle, Bodie, Dolgas, & Herman, 2003). The mPFC indirectly provides negative feedback to the HPA axis (Diorio, Viau, & Meaney, 1993; Radley, Arias, & Sawchenko, 2006; Spencer, Buller, & Day, 2005) and modulates fear responses (Diorio et al., 1993; Radley et al., 2006; Sapolsky et al., 1984). Prelimbic (PrL) and Infralimbic (IL) regions are the two central sub-regions that form mPFC. PrL is shown to play a role in fear-conditioning and learning while fear extinction is associated with IL (Maren & Quirk, 2004). mPFC region is sensitive to GC, which regulates stress response in a biphasic manner. The GR:MR ratio in mPFC is high which reflects the high traffic of stress hormones in this region (Joëls, Sarabdjitsingh, & Karst, 2012). Hence, the primary brain regions regulating HPA axis include the hippocampus, the amygdala and the medial prefrontal cortex (Brain-HPA).

Therefore, the interaction of the brain Hippocampus-PFC-Amygdala axes and Hypothalamus-Pituitary-Adrenal axes influences the behavioral readout of an individual (Fig. 1). This double-HPA nestled within us is the core of our physiological, behavioral and stress-response regulation. As seen earlier, stress has significant impacts on the *structure of brain, neuroendocrine system* and *behavioral outcome*, which are the interacting triad in response to stress.

5. Influences of differential early-life environment on brain function?

The environment plays a crucial role in molding BLA, hippocampus, and mPFC, the regions of the brain that are involved in regulating the HPA axis as mentioned above. Social attachment is an integral environmental component that influences an individual's behavior. The "social brain hypothesis" explains the large brain size of primates. It states that a primate's social group size has a positive correlation with the volume of its prefrontal cortex (Dunbar, 1998). This correlation between PFC volume and social network size is found in humans as well, which is reflected in their social cognitive abilities (Powell, Lewis, Roberts, Garcia-Finana, & Dunbar, 2012).

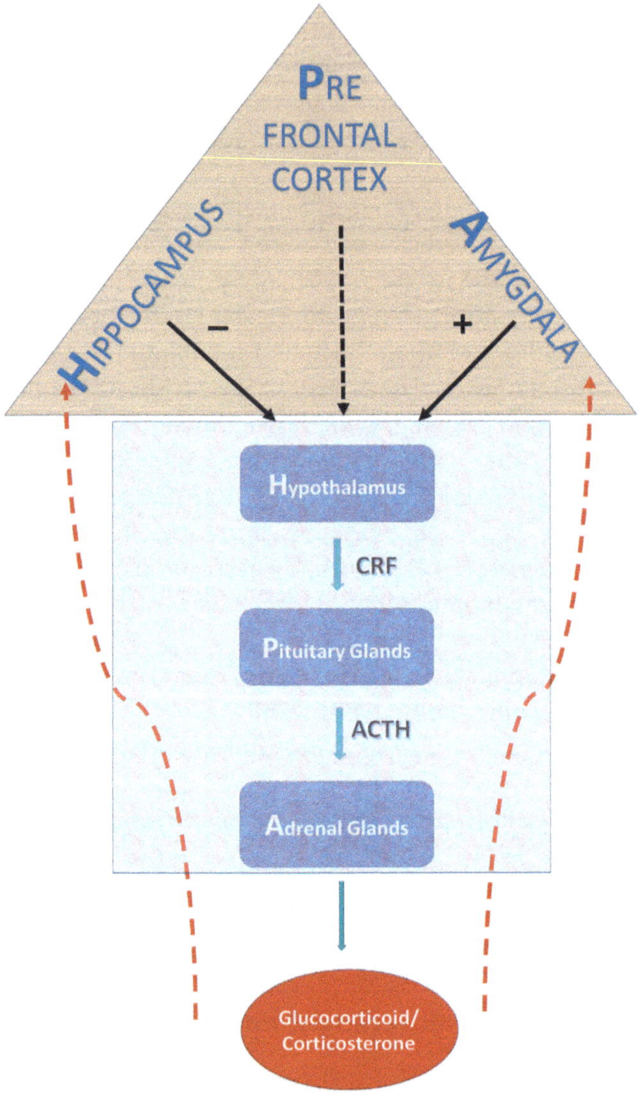

Fig. 1 The "double-HPA"-home of physiological regulation, involving the central HPA (within brain, schematically presented as the home-roof) and the systemic HPA (within our body, schematically presented as the home-block). Central HPA involves Hippocampus, Prefrontal cortex and Amygdala each of which influence (#) stress-HPA, involving hypothalamus, pituitary and adrenal glands to secrete glucocorticoids (GC) or corticosterone (show in blue block arrows). GC in turn influences the Amygdala and the Hippocampus through binding to its receptors (show in dashed red arrows). CRF: corticotrophin release factor and ACTH: adrenocorticotrophic hormone regulates the stress-HPA axis. # The hippocampus negatively regulates hypothalamus-pituitary-adrenal axis (black solid arrow with negative sign), while the amygdala positively regulates the same (black solid arrow with positive sign). The prefrontal cortex indirectly regulates the hypothalamus-pituitary-adrenal axis (black dashed arrow).

In a recent study complexity of environment has been shown to have direct impact on complexity of mPFC neurons, showing social environment's foothold on cortical plasticity (Ashokan, Lim, Hang, & Mitra, 2018). This shows that the environment plays a significant role in modulating brain regions.

As mentioned earlier, maternal care is a crucial need of an individual during early life, as it helps in the regulation of the HPA axis and emotional stability in the long run (Hall, 2014; Zalewski, Lengua, Kiff, & Fisher, 2012). Orphans growing up in institutions deprived of social attachment during their early life display physical, psychological, and sexual aggression (Boucher, Pare, Perry, Sigal, & Ouimet, 2008). Impacts of mother and infant attachment were studied in the 1950s in a famous experiment on Rhesus Monkeys by a psychologist, Harry Harlow. He showed that the mother-infant relationship was more psychological than biological. The study suggested that mothers were important not just for supplementing food but also crucial for touch-sensitive attachment with the infant that helped the long-term emotional and mental stability of the child (Harlow & Zimmermann, 1959).

In contrast, positive influences in early life show a healthy pattern of development in an individual. The impact of strong parental bonding, characterized by high affection toward children, is linked to a lower risk of PTSD in adulthood, suggesting that emotional enrichment protects against PTSD in adulthood (Lima et al., 2014). Contact behavior is particularly crucial for mother-infant interaction. Besides parental touch, face-to-face interaction is vital for infants to be an active participant in social encounters during their growing years (Mantis, Stack, Ng, Serbin, & Schwartzman, 2014). Another excellent example of an early life environment promoting learning and social interactions is speech. New evidence reveals that starting from the first months of life, listening to speech is a reliable tool to support the attainment of fundamental psychological processes like pattern learning, categorizing the formation of objects, identifying partners for communication, improving social interactions, and developing social cognition (Vouloumanos & Waxman, 2014).

6. Environmental control of stress resilience

We need to address the important point of developing interventions against concerns associated with early life stress. Individuals vary in reacting to an adverse situation wherein not all individuals react in a similar manner toward adversity. When adults or children are exposed to a stressful environment, most of them are maladaptive, while few of the individuals respond in a well adaptive manner and are resilient (Rutter, 1985). Therefore, processes and mechanisms driving resilience need to be studied thoroughly so that these mechanisms can be utilized as an intervention to all the detrimental effects caused by early life stress (Jackson, Firtko, & Edenborough, 2007). The adults who are resilient (well adaptive) toward a stressful episode might have been exposed to a stimulatory and enriched environment during early life or a mixture of stress and enriched environment during early life.

To define and understand the neurobiology of resilience, we need to focus on the other side of the coin, "stress," from where different coping strategies/resilience starts. Maternal separation is a well-established early life stress model (Huot, Gonzalez, Ladd, Thrivikraman, & Plotsky, 2004), and Enriched environment (EE) is a paradigm that induces pro-resilient behavioral and physiological features (Ashokan et al., 2016; Francis, Diorio, Plotsky, & Meaney, 2002) as well as adaptive behavioral and neural features in animals (Mohammed et al., 2002; van Praag, Kempermann, & Gage, 2000). Several studies have determined that an enriched environment can rescue the damaging effects of early life stress. Environmental enrichment during the peri-pubertal period rescues the harmful effects of maternal separation on HPA and behavioral response to stressors (Francis et al., 2002). Short term enrichment reverses the anxiety and hypertrophy of BLA neurons in adulthood (Koe, Ashokan, & Mitra, 2016). Since there is such limited information about how enriched environment can be applied in early life to counter stress and promote resilience, it makes it a fascinating area to explore.

Animal models are invaluable for an understanding of the mechanisms involved in the sequelae of early life environment. An interesting early-life study in rodents showed that natural fear of predator odor develops after postnatal day (PND) 10, and this is correlated with the function of the amygdala and regulation of corticosterone (CORT) (Moriceau, Roth, Okotoghaide, & Sullivan, 2004). As mentioned in the background, the study highlighted that maternal presence acts as an environmental switch during infancy for stabilizing fear responses (Moriceau & Sullivan, 2006). At this point we know that harmful effects of MS can be reversed via short-term enrichment in adulthood (Koe et al., 2016). The following paragraphs will discuss the background of the two crucial paradigms of MS and EE.

6.1 Maternal separation (MS)

Various studies have shown that any form of alteration in maternal care can have a lasting effect on the stress response, behavior, and emotion in rats, as it is one of the most crucial influences during development (Champagne, Francis, Mar, & Meaney, 2003). Manipulations in the early environment cause acute and long-term effects on endocrinology and behavior, which have been studied in various models, including rats, mice, primates, etc. Bonnet infant monkeys show a compromised behavioral and physiological response to maternal separation. Similarly, pig-tailed and rhesus infant monkeys exhibit a maladaptive response to maternal separation wherein the mother and the conspecifics are removed (Bailey & Coe, 1999; Laudenslager, Reite, & Harbeck, 1982; Reite, Kaemingk, & Boccia, 1989; Reite, Short, Kaufman, Stynes, & Pauley, 1978).

In rats, one of the most well-studied environmental manipulations that have been reported is a brief daily separation of the pups from the mother (dam) before and after weaning. Levine et al. introduced a protocol where they separated the rat pups from dams for only 15 min, which was typically referred to as "early handling" (EH) but still considered a short maternal deprivation (Levine, 1956). In the mid-1970s a different but related procedure was introduced, namely that of prolonged maternal separation (MS). Maternal separation is a collective term used to describe various experimental methods that involve separation of a pup from the dam for different periods and is an undoubtedly promising model to study early life stress.

A large body of evidence shows that MS during postnatal weeks causes a permanent increase in anxiety-related disorders (Francis, Champagne, Liu, & Meaney, 1999). Lifelong hyperactivity of the HPA (hypothalamus-pituitary-adrenal) axis is another consequence of this postnatal separation (Huot, Plotsky, Lenox, & McNamara, 2002). Maternal care involving licking/grooming (LG) and arched-back nursing (ABN) of their pups (LG-ABN) is a strong influence on early life development, as maternal care is the critical environment that molds the pup's long-lasting coping with stress. Such maternal behaviors play a decisive role in regulating the HPA axis of the pup (Fish et al., 2004), and these high LG offspring exhibit similar maternal behavior later on (Hellstrom, Dhir, Diorio, & Meaney, 2012).

It has been found that when compared to control rats, MS rats show depressive-like behavior in the forced swimming test in adulthood (Plotsky, Owens, & Nemeroff, 1998) and anxiety-like behavior in the elevated plus-maze (Huot, Thrivikraman, Meaney, & Plotsky, 2001).

MS rats also have increased HPA axis responsiveness to stressors and elevated CRF mRNA in the PVN even in later stages of life (Plotsky & Meaney, 1993). As stated earlier, hippocampal GR causes glucocorticoid negative feedback, and hippocampal GR loss causes elevated GC levels. Therefore, there is a decrease in inhibitory signaling to the PVN-CRF neurons due to the downregulation of hippocampal GR (Sapolsky et al., 1986). Apart from anxiety, MS affects cognitive and social behaviors too (Sabatini et al., 2007; Seay, Alexander, & Harlow, 1964). A study on the behavior of MS rats exhibited a significant cognitive impairment in two different tests, the Morris water maze task and the novel object recognition test (Huot et al., 2002; Lehmann, Pryce, Bettschen, & Feldon, 1999). Therefore, neonatal MS in rats can be considered as a suitable animal model for stress-vulnerable organisms during early life that resulted in depression-like syndrome and enhanced stress responsiveness (Aisa, Tordera, Lasheras, Del Rio, & Ramirez, 2007).

6.2 Effects of MS on neuronal plasticity and molecules in the brain

Early-life stress leads to reductions in hippocampal neurogenesis as well as retraction of dendrites and spine density of the hippocampus, medial prefrontal cortex (mPFC), and nucleus accumbens (Monroy et al., 2010). A single episode of MS for 24 h at postnatal day 3 does not affect dendritic complexity in the basolateral amygdala (BLA) (Krugers et al., 2012). 3 h of MS from P2-P14 caused hypertrophy of BLA neurons, which was observed both in dendritic length and branch points (Koe et al., 2016). These are the only two studies that give us information about the BLA neuronal morphology with respect to the MS paradigm. This limited information available about MS and BLA makes it intriguing to explore further.

Various research using rodent models provides evidence for the role of early life stress on the epigenome. Poor maternal care in rats alters DNA methylation at a particular sequence of the glucocorticoid receptor gene (Nr3c1) in the hippocampus of the pup, causing epigenetic and behavioral changes in adulthood. MS caused the downregulation of hippocampal GR (Weaver et al., 2004). However, lack of epigenetic changes has also been reported in MS rats where DNA methylation in the same region of Nr3c1 of the hippocampus does not occur (Daniels et al., 2009). MS rats are seen to have long-term adverse effects like increased levels of plasma ACTH. The effects of maternal deprivation were not only limited to the HPA axis, but elevated CRH mRNA expression was also seen in the amygdala, which most certainly plays a role in the stress response (Anisman et al., 1998). This shows the influence of molecules in the amygdala mediating stress response in MS rats. MS rats exposed to acute restraint stress were reported to have elevated levels of recently active neurons in the central nucleus of the amygdala, PVN, and bed nucleus of the stria terminalis, brain regions that play a critical role in organizing the behavioral response to stress (Day, Nebel, Sasse, & Campeau, 2005; Koehnle & Rinaman, 2010; Sanders & Anticevic, 2007). Though very little is known about the CRF receptors, a few studies have shown that after MS for 24 h CRF receptor two is down-regulated in ventromedial hypothalamus and biomedical amygdala (Rivarola & Renard, 2014). Prior studies show that global effect of early life experience on the hippocampal transcriptome alters more than 900 genes in response to low maternal care that in turn causes anxiety-like behavior. A few of these changes seen in gene expression are reversed using histone deacetylase inhibitor and methyl donor. Such epigenetic changes due to MS have been widely studied in the hippocampus (Daniels et al., 2009; Moffett et al., 2007; Weaver, Meaney, & Szyf, 2006).

6.3 Enriched environment (EE)

Though the positive effects of environmental stimulation and its enrichment have been studied for many years, it was Hebb in the late 1940s who proposed the "enriched environment" as an experimental concept. He reported that his pet rats displayed behavioral improvements as compared to his laboratory rats (Hebb, 1947). In 1942, Robert Tyron raised two groups of "bright" and "dull" mice by repeatedly breeding a "bright" male (few mistakes made in the Hebb-William maze test) with a "bright" female and "dull" male (multiple mistakes made in the Hebb-William maze test) with a "dull" female. He took seven generations to achieve a state where the resultant progenies of both the lines were all "bright" and "dull," respectively (Tryon, 1940). Following the same line of research, Cooper and Zubek (1958) exposed a group of "bright" and "dull" to an enriched environment (spacious and stimulatory living conditions) and another group to the standard housing to compare the effects of housing on the rat's behavior. This led to an interesting result where EE elevated the learning capacity in "dull" rats while EE did not affect "bright" rats. The standard conditions alleviated the learning ability in "bright" mice and did not influence "dull" rats. The ability of a simple environment to change the phenotype in one generation that took seven generations of inbreeding previously is an outstanding observation (Cooper & Zubek, 1958). Modern experimental approaches build on this by using an experimental "enriched" environment, where the cages were much larger with more complex social interaction, and it also consisted of the exploratory environment with tunnels, toys, nesting material, and running wheels in some cases (Brown et al., 2003; Kempermann, Kuhn, & Gage, 1997; Komitova, Zhao, Gidö, Johansson, & Eriksson, 2005). Some studies exclude voluntary running wheel to separate sensory complexity from effects of the physical exercise (Ashokan et al., 2016; Kobilo et al., 2011; Mustroph et al., 2012).

Mice exposed to the enriched environment have enhanced memory function in various learning tasks. These mice perform better in a spatial memory task than controls in standard housing cage (Pacteau, Einon, & Sinden, 1989). Rodents exposed to an enriched environment, both before and after a cerebral insult, show improved cognitive performance (Young, Lawlor, Leone, Dragunow, & During, 1999). Recent studies have demonstrated that EE can influence innate behaviors like avoidance of predator odor and attractiveness of males (Mitra & Sapolsky, 2012). Clinical evidence shows a reduced risk of dementia in individuals with Parkinson's and Alzheimer's disease exposed to a stimulating environment in addition to positive effects on cerebral health (Young et al., 1999). Enrichment is also seen to reverse the effects of maternal separation on the HPA axis, where the MS rats are subjected to EE during the peripubertal stage. When the plasma corticosterone levels are compared between the MS and MS-EE rats, the latter show much lower corticosterone levels than the MS rats (Francis et al., 2002).

6.4 Effects of EE on neuronal plasticity and molecules in the brain

The cerebral cortex is the area that takes part in higher cognitive processing, and this region is more receptive than other areas of the brain to environmental enrichment. The cortex of animals exposed to environmental enrichment show an increase in thickness compared with the standard group, while the brains from the impoverished group show a decrease in cortical thickness (van Praag et al., 2000). These changes are thought to be due to several factors, including increased soma size, number, and length of dendrites, dendritic spines, and length of postsynaptic thickening as measured on electron microscopic pictures of synapses (Diamond, 2001; Diamond, Krech, & Rosenzweig, 1964).

Environmental enrichment and voluntary exercise exhibit similar effects on the brain. Studies showed that this change in environment and playing on wheels affected neurogenesis. Running enhances the survival of the newborn neurons in the dentate gyrus (Faverjon et al., 2002; van Praag, Christie, Sejnowski, & Gage, 1999). In addition to its effects on neurogenesis, studies have shown that an enriched environment reduces spontaneous apoptotic cell death in a rodent's hippocampus by 45% (Young et al., 1999). Moreover, these environmental conditions have a protective effect against kainate-induced seizures and excitotoxic injury.

The first molecular player to focus on the HPA axis is the glucocorticoid receptor (GR), GC activated transcription factors that maintain the regulation of HPA axis in response to various stressors (Furay, Bruestle, & Herman, 2008; Herman & Cullinan, 1997; Joëls et al., 2012). The density of the GR population across the amygdala, hippocampus, and mPFC differ and have a link in providing feedback regulation via the HPA axis (Herman & Cullinan, 1997). The recruitment of activated GR in the nucleus as a transcription factor is a crucial step that initiates a relay of downstream molecular pathways in various brain regions to mediate the stress response (Bamberger et al., 1996; Hollenberg & Evans, 1988). Prior studies show that activation of GR through GC causes phosphorylation of Erk1/2^{MAPK}. This activated Erk1/2^{MAPK} causes relay of downstream molecular pathways that are crucial in mediating behavioral effects of GC (Jean-Michel et al., 2005; Revest et al., 2005). Another critical molecule that interacts with GC-GR and Erk1/2^{MAPK} is BDNF that mediates molecular cascade to enhance fear-related memories. The enrichment also induces expression of glial-derived neurotrophic factor (GDNF) and brain-derived neurotrophic factor (BDNF) (Nithianantharajah & Hannan, 2006; Young et al., 1999). It also increases phosphorylation of the transcription factor cyclic-AMP response element-binding protein, indicating that the influence of the enriched environment on spontaneous apoptosis may be mediated through transcription factor activation (Young et al., 1999).

Similarly, ischemic injury within the brain decreases nerve growth factor-induced gene A (NGFI-A) mRNA expression in cortical areas and the CA1 subregion of the hippocampus two to 3 days after insult. However, when these rats are housed in an enriched environment, NGFI-A expression increases compared to standard housing, and the stroke-induced reduction of the hippocampal glucocorticoid receptor (GR) mRNA is reverted. By altering these molecules after the stroke, an improved functional outcome can be attained by exposure to long-term EE (Dahlqvist et al., 1999;

Johansson, 1996; van Praag et al., 2000). A lot of focus in research is directed toward the response of EE in the hippocampus in a broad range of biological variables. The neurotrophic factors like nerve growth factor (NGF), brain-derived neurotrophic factor (BDNF), and neurotrophin-3 (NT-3) are some of the significant factors to regulate the dendritic growth and the arborization of the pyramidal neurons in the hippocampus. NGF regulates structural plasticity in neurons and is upregulated under an enriched environment in the hippocampus to give rise to increased dendritic arborization and spine density (Sofroniew, Howe, & Mobley, 2001). NT-3 mRNA is also upregulated in the hippocampus of enriched rats. BDNF has a crucial role in modulating apical dendritic spine density and the number of spines. This neurotrophic factor also has a high expression in hippocampus in response to EE (Falkenberg et al., 1992).

While several studies have provided EE in the adulthood, relatively fewer studies incorporate EE in the early stages of the life. The available pieces of the work in the early-life EE includes a conspicuous acceleration of visual system development (Sale et al., 2004), reduction in neurochemical deficits of Huntington's disease (Glass, Van Dellen, Blakemore, Hannan, & Faull, 2004), prevention of the prenatal stress-induced cognitive disorders (Yang et al., 2007) and rescue of the detrimental effects of limited nesting on young adult rats (Cui et al., 2006).

7. Conclusion

The resilience to stress is a critical aspect of health across the lifespan. Hence, research pertaining to the neurobiology of stress resilience will continue to be important in the future. Central regulation of stress hormones by the brain and allostatic changes in how the brain recognizes stress hormones will hopefully provide prognostic tools leading to healthier lifestyles. Additionally, the reactivity of stressors is a function of the personal history of prior stress exposure. Thus, the tailoring of psychological therapy and pharmacological treatment to the personalized experience is also a critical area for future investigations. Closer integration of the environmental influences on brain and behavior into stress susceptibility or resilience is thus an important area for future research.

Living environment robustly affects our mental well-being. Conclusions about stress susceptibility or resilience on mental health can be drawn only after analyzing environmental conditions influencing them. For example, there is an increasing realization that urban living environment causes a significant psychological burden on its inhabitants. Given the increasing pace of urbanization, this is an urgent imperative for policymakers and healthcare providers to substantiate their contribution to mental health awareness. Robust effects of environmental enrichment in animals suggest that a similar species-typical enrichment of humans can buffer the adverse consequences of urbanization. Thus, the provision of green spaces in a city is not merely an aesthetic addendum but it also has a health imperative of substantial economic and societal value. Similarly, for long, the relationship between our living environment and human health into it has been tackled with a monolithic vision of socioeconomics. We now know that the socioeconomic gradient can predispose those on lower rungs to diminished health even when access to healthcare is available. Hence, safe neighborhoods and nurturing housing environment is not only a social construct but also a palpable biological necessity.

The increasing focus of neurobiological research on early life is timely because of the importance of environmental conditions during this critical and sensitive phase in human and animal life. The formative period of maternal interactions can have immense consequences for later physiology and behavior of children. As discussed here, there are innumerable examples of the negative effects of a poor nurturing environment in early-life. On the other side, ensuring an enriching environment can provide a buffer to the challenges brought upon both the mother and the child during this demanding phase. In other words, a happy maternal life makes happier offspring for the future.

Acknowledgment

This work is supported by RG 144/17, Ministry of Education, Singapore.

References

Ainsworth, M. S. (1979). Infant–mother attachment. *American Psychologist, 34*(10), 932.

Ainsworth, M. D. S., & Bell, S. M. (1970). Attachment, exploration, and separation: Illustrated by the behavior of one-year-olds in a strange situation. *Child Development, 41*, 49–67.

Aisa, B., Tordera, R., Lasheras, B., Del Rio, J., & Ramirez, M. J. (2007). Cognitive impairment associated to HPA axis hyperactivity after maternal separation in rats. *Psychoneuroendocrinology, 32*(3), 256–266. https://doi.org/10.1016/j.psyneuen.2006.12.013.

Allen, J., & Allen, C. (1974). Role of the amygdaloid complexes in the stress-induced release of ACTH in the rat. *Neuroendocrinology, 15*(3–4), 220–230.

Allen, J., & Allen, C. (1975). Amygdalar participation in tonic ACTH secretion in the rat. *Neuroendocrinology, 19*(2), 115–125.

Anisman, H., Zaharia, M. D., Meaney, M. J., & Merali, Z. (1998). Do early-life events permanently alter behavioral and hormonal responses to stressors? *International Journal of Developmental Neuroscience*, *16*(3–4), 149–164.

Ashokan, A., Hegde, A., & Mitra, R. (2016). Short-term environmental enrichment is sufficient to counter stress-induced anxiety and associated structural and molecular plasticity in basolateral amygdala. *Psychoneuroendocrinology*, *69*, 189–196. https://doi.org/10.1016/j.psyneuen.2016.04.009.

Ashokan, A., Lim, J. W. H., Hang, N., & Mitra, R. (2018). Complex housing causes a robust increase in dendritic complexity and spine density of medial prefrontal cortical neurons. *Scientific Reports*, *8*(1), 7308. https://doi.org/10.1038/s41598-018-25399-4.

Bailey, M. T., & Coe, C. L. (1999). Maternal separation disrupts the integrity of the intestinal microflora in infant rhesus monkeys. *Developmental Psychobiology*, *35*(2), 146–155.

Bamberger, C. M., Schulte, H. M., & Chrousos, G. P. (1996). Molecular determinants of glucocorticoid receptor function and tissue sensitivity to glucocorticoids. *Endocrine Reviews*, *17*(3), 245–261.

Barker, D. J. P. (1998). *Mothers, babies and health in later life.* Elsevier Health Sciences.

Barker, D. J., Eriksson, J. G., Forsén, T., & Osmond, C. (2002). Fetal origins of adult disease: Strength of effects and biological basis. *International Journal of Epidemiology*, *31*(6), 1235–1239.

Bateson, P. (2001). Fetal experience and good adult design. *International Journal of Epidemiology*, *30*(5), 928–934.

Baydar, N., & Brooks-Gunn, J. (1991). Effects of maternal employment and child-care arrangements on preschoolers' cognitive and behavioral outcomes: Evidence from the children of the National Longitudinal Survey of youth. *Developmental Psychology*, *27*(6), 932.

Berton, O., Covington, H. E., Ebner, K., Tsankova, N. M., Carle, T. L., Ulery, P., et al. (2007). Induction of ΔFosB in the periaqueductal gray by stress promotes active coping responses. *Neuron*, *55*(2), 289–300.

Bohlin, G., Hagekull, B., & Rydell, A. M. (2000). Attachment and social functioning: A longitudinal study from infancy to middle childhood. *Social Development*, *9*(1), 24–39.

Boucher, S., Pare, N., Perry, J. C., Sigal, J. J., & Ouimet, M. C. (2008). Consequences of an institutionalized childhood: The case of the "Duplessis orphans" *Santé Mentale au Québec*, *33*(2), 271–291.

Bowlby, J. (1958). The nature of the child's tie to his mother. *The International Journal of Psycho-Analysis*, *39*, 350.

Bradley, R. H., Whiteside, L., Mundfrom, D. J., Casey, P. H., Kelleher, K. J., & Pope, S. K. (1994). Early indications of resilience and their relation to experiences in the home environments of low birthweight, premature children living in poverty. *Child Development*, *65*(2), 346–360.

Brown, J., Cooper-Kuhn, C. M., Kempermann, G., Van Praag, H., Winkler, J., Gage, F. H., et al. (2003). Enriched environment and physical activity stimulate hippocampal but not olfactory bulb neurogenesis. *European Journal of Neuroscience*, *17*(10), 2042–2046.

Cameron, N. M., Shahrokh, D., Del Corpo, A., Dhir, S. K., Szyf, M., Champagne, F. A., et al. (2008). Epigenetic programming of phenotypic variations in reproductive strategies in the rat through maternal care. *Journal of Neuroendocrinology*, *20*(6), 795–801. https://doi.org/10.1111/j.1365-2826.2008.01725.x.

Walter Bradford Cannon. (1932a). *Effects of strong emotions.* University of Chicago Press.

Walter Bradford Cannon. (1932b). Homeostasis. In *The wisdom of the body.* New York: Norton.

Champagne, F. A., Francis, D. D., Mar, A., & Meaney, M. J. (2003). Variations in maternal care in the rat as a mediating influence for the effects of environment on development. *Physiology & Behavior*, *79*(3), 359–371.

Cicchetti, D., & Rogosch, F. A. (2009). Adaptive coping under conditions of extreme stress: Multilevel influences on the determinants of resilience in maltreated children. *New Directions for Child and Adolescent Development, 2009*(124), 47–59.
Cohen, S., Janicki-Deverts, D., & Miller, G. E. (2007). Psychological stress and disease. *JAMA, 298*(14), 1685–1687. https://doi.org/10.1001/jama.298.14.1685.
Cooper, R. M., & Zubek, J. P. (1958). Effects of enriched and restricted early environments on the learning ability of bright and dull rats. *Canadian journal of psychology, 12*(3), 159.
Cui, M., Yang, Y., Yang, J., Zhang, J., Han, H., Ma, W., et al. (2006). Enriched environment experience overcomes the memory deficits and depressive-like behavior induced by early life stress. *Neuroscience Letters, 404*(1–2), 208–212.
Dahlqvist, P., Zhao, L., Johansson, I. M., Mattsson, B., Johansson, B. B., Seckl, J. R., et al. (1999). Environmental enrichment alters nerve growth factor-induced gene A and glucocorticoid receptor messenger RNA expression after middle cerebral artery occlusion in rats. *Neuroscience, 93*(2), 527–535.
Daniels, W., Fairbairn, L., Van Tilburg, G., McEvoy, C., Zigmond, M., Russell, V., et al. (2009). Maternal separation alters nerve growth factor and corticosterone levels but not the DNA methylation status of the exon 17 glucocorticoid receptor promoter region. *Metabolic Brain Disease, 24*(4), 615.
Dantzer, B., Newman, A. E., Boonstra, R., Palme, R., Boutin, S., Humphries, M. M., et al. (2013). Density triggers maternal hormones that increase adaptive offspring growth in a wild mammal. *Science, 340*(6137), 1215–1217. https://doi.org/10.1126/science.1235765.
Davis, M. (1992). The role of the amygdala in fear and anxiety. *Annual Review of Neuroscience, 15*(1), 353–375. https://doi.org/10.1146/annurev.ne.15.030192.002033.
Davis, M. (1998). Are different parts of the extended amygdala involved in fear versus anxiety? *Biological Psychiatry, 44*(12), 1239–1247.
Davis, M., Rainnie, D., & Cassell, M. (1994). Neurotransmission in the rat amygdala related to fear and anxiety. *Trends in Neurosciences, 17*(5), 208–214.
Day, H. E., Nebel, S., Sasse, S., & Campeau, S. (2005). Inhibition of the central extended amygdala by loud noise and restraint stress. *The European Journal of Neuroscience, 21*(2), 441–454. https://doi.org/10.1111/j.1460-9568.2005.03865.x.
Desai, M., & Hales, C. N. (1997). Role of fetal and infant growth in programming metabolism in later life. *Biological Reviews, 72*(2), 329–348.
Diamond, M. C. (2001). Response of the brain to enrichment. *Anais da Academia Brasileira de Ciências, 73*(2), 211–220.
Diamond, M. C., Krech, D., & Rosenzweig, M. R. (1964). The effects of an enriched environment on the histology of the rat cerebral cortex. *The Journal of comparative neurology, 123*, 111–120.
Diorio, D., Viau, V., & Meaney, M. J. (1993). The role of the medial prefrontal cortex (cingulate gyrus) in the regulation of hypothalamic-pituitary-adrenal responses to stress. *Journal of Neuroscience, 13*(9), 3839–3847.
Dunbar, R. (1998). The social brain hypothesis. *Brain*, 9(10), 178–190.
Elias, M. J., & Arnold, H. (2006). *The educator's guide to emotional intelligence and academic achievement: Social-emotional learning in the classroom.* Corwin Press.
Eriksson, J. G., Forsen, T., Tuomilehto, J., Winter, P. D., Osmond, C., & Barker, D. J. (1999). Catch-up growth in childhood and death from coronary heart disease: Longitudinal study. *BMJ, 318*(7181), 427–431.
Falkenberg, T., Mohammed, A. K., Henriksson, B., Persson, H., Winblad, B., & Lindefors, N. (1992). Increased expression of brain-derived neurotrophic factor mRNA in rat hippocampus is associated with improved spatial memory and enriched environment. *Neuroscience Letters, 138*(1), 153–156.

Faulkner, B., & Brown, T. H. (1999). Morphology and physiology of neurons in the rat perirhinal-lateral amygdala area. *The Journal of comparative neurology*, *411*(4), 613–642.

Faverjon, S., Silveira, D., Fu, D., Cha, B., Akman, C., Hu, Y., et al. (2002). Beneficial effects of enriched environment following status epilepticus in immature rats. *Neurology*, *59*(9), 1356–1364.

Field, T., Healy, B. T., Goldstein, S., & Guthertz, M. (1990). Behavior-state matching and synchrony in mother-infant interactions of nondepressed versus depressed dyads. *Developmental Psychology*, *26*(1), 7.

Figueiredo, H. F., Bruestle, A., Bodie, B., Dolgas, C. M., & Herman, J. P. (2003). The medial prefrontal cortex differentially regulates stress-induced c-fos expression in the forebrain depending on type of stressor. *European Journal of Neuroscience*, *18*(8), 2357–2364.

Fish, E. W., Shahrokh, D., Bagot, R., Caldji, C., Bredy, T., Szyf, M., et al. (2004). Epigenetic programming of stress responses through variations in maternal care. *Annals of the New York Academy of Sciences*, *1036*(1), 167–180.

Francis, D. D., Champagne, F. A., Liu, D., & Meaney, M. J. (1999). Maternal care, gene expression, and the development of individual differences in stress reactivity. *Annals of the New York Academy of Sciences*, *896*(1), 66–84.

Francis, D. D., Diorio, J., Plotsky, P. M., & Meaney, M. J. (2002). Environmental enrichment reverses the effects of maternal separation on stress reactivity. *The Journal of Neuroscience*, *22*(18), 7840–7843.

Furay, A. R., Bruestle, A. E., & Herman, J. P. (2008). The role of the forebrain glucocorticoid receptor in acute and chronic stress. *Endocrinology*, *149*(11), 5482–5490.

Garland, T., Jr., & Kelly, S. A. (2006). Phenotypic plasticity and experimental evolution. *The Journal of Experimental Biology*, *209*(Pt. 12), 2344–2361. https://doi.org/10.1242/jeb.02244.

Gilbert, S. F. (2001). Ecological developmental biology: Developmental biology meets the real world. *Developmental Biology*, *233*(1), 1–12.

Gilbert, S. F. (2004). Mechanisms for the environmental regulation of gene expression. *Birth Defects Res. C Embryo Today*, *72*(4), 291–299.

Glass, M., Van Dellen, A., Blakemore, C., Hannan, A., & Faull, R. (2004). Delayed onset of huntington' s disease in mice in an enriched environment correlates with delayed loss of cannabinoid CB1 receptors. *Neuroscience*, *123*(1), 207–212.

Gluckman, P. D., & Hanson, M. A. (2004). The developmental origins of the metabolic syndrome. *Trends in endocrinology and metabolism: TEM*, *15*(4), 183–187.

Godfrey, K., Barker, D., Robinson, S., & Osmond, C. (1997). Maternal birthweight and diet in pregnancy in relation to the infant's thinness at birth. *BJOG: An International Journal of Obstetrics & Gynaecology*, *104*(6), 663–667.

Groothuis, T. G. G., Müller, W., von Engelhardt, N., Carere, C., & Eising, C. (2005). Maternal hormones as a tool to adjust offspring phenotype in avian species. *Neuroscience & Biobehavioral Reviews*, *29*(2), 329–352. https://doi.org/10.1016/j.neubiorev.2004.12.002.

Hall, A. W. M. (2014). *Development of emotion regulation neural circuitry: Anatomical volumes and functional connectivity in middle childhood,* MSc thesis, McMaster University.

Harlow, H. F., & Zimmermann, R. R. (1959). Affectional responses in the infant monkey; orphaned baby monkeys develop a strong and persistent attachment to inanimate surrogate mothers. *Science*, *130*(3373), 421–432.

Hebb, D. O. (1947). The effects of early experience on problem solving at maturity. *The American Psychologist*, *2*, 306–307.

Hellstrom, I. C., Dhir, S. K., Diorio, J. C., & Meaney, M. J. (2012). Maternal licking regulates hippocampal glucocorticoid receptor transcription through a thyroid hormone–serotonin–NGFI-A signalling cascade. *Philosophical Transactions of the Royal Society, B: Biological Sciences*, *367*(1601), 2495–2510.

Henry, C. J. K., & Ulijaszek, S. J. (1996). Long-term consequences of early environment: Growth, development and the lifespan developmental perspective. *Vol. 37*. Cambridge University Press.

Herman, J. P., & Cullinan, W. E. (1997). Neurocircuitry of stress: Central control of the hypothalamo–pituitary–adrenocortical axis. *Trends in Neurosciences*, *20*(2), 78–84.

Herman, J. P., Prewitt, C. M.-F., & Cullinan, W. E. (1996). Neuronal circuit regulation of the hypothalamo-pituitary-adrenocortical stress axis. *Critical Reviews in Neurobiology*, *10*(3-4), 371–394.

Herman, J. P., Schafer, M., Young, E. A., Thompson, R., Douglass, J., Akil, H., et al. (1989). Evidence for hippocampal regulation of neuroendocrine neurons of the hypothalamo-pituitary-adrenocortical axis. *Journal of Neuroscience*, *9*(9), 3072–3082.

Hollenberg, S. M., & Evans, R. M. (1988). Multiple and cooperative trans-activation domains of the human glucocorticoid receptor. *Cell*, *55*(5), 899–906.

Huot, R. L., Gonzalez, M. E., Ladd, C. O., Thrivikraman, K. V., & Plotsky, P. M. (2004). Foster litters prevent hypothalamic-pituitary-adrenal axis sensitization mediated by neonatal maternal separation. *Psychoneuroendocrinology*, *29*(2), 279–289. https://doi.org/10.1016/S0306-4530(03)00028-3.

Huot, R. L., Plotsky, P. M., Lenox, R. H., & McNamara, R. K. (2002). Neonatal maternal separation reduces hippocampal mossy fiber density in adult long Evans rats. *Brain Research*, *950*(1), 52–63.

Huot, R. L., Thrivikraman, K., Meaney, M. J., & Plotsky, P. M. (2001). Development of adult ethanol preference and anxiety as a consequence of neonatal maternal separation in long Evans rats and reversal with antidepressant treatment. *Psychopharmacology*, *158*(4), 366–373.

Issa, A. M., Rowe, W., Gauthier, S., & Meaney, M. J. (1990). Hypothalamic-pituitary-adrenal activity in aged, cognitively impaired and cognitively unimpaired rats. *Journal of Neuroscience*, *10*(10), 3247–3254.

Jackson, D., Firtko, A., & Edenborough, M. (2007). Personal resilience as a strategy for surviving and thriving in the face of workplace adversity: A literature review. *Journal of Advanced Nursing*, *60*(1), 1–9.

Jean-Michel, R., Di Blasi, F., Kitchener, P., Aline, D., Marc, T., & Piazza, P. V. (2005). Erratum: The MAPK pathway and Egr-1 mediate stress-related behavioral effects of glucocorticoids. *Nature Neuroscience*, *8*(6), 835.

Joëls, M., Sarabdjitsingh, R. A., & Karst, H. (2012). Unraveling the time domains of corticosteroid hormone influences on brain activity: Rapid, slow, and chronic modes. *Pharmacological Reviews*, *64*(4), 901–938.

Johansson, B. B. (1996). Functional outcome in rats transferred to an enriched environment 15 days after focal brain ischemia. *Stroke*, *27*(2), 324–326.

Kan, J. M., Callaghan, B. L., & Richardson, R. (2016). A mother's past can predict her offspring's future: Previous maternal separation leads to the early emergence of adult-like fear behavior in subsequent male infant rat offspring. *Behavioral Neuroscience*, *130*, 511–520.

Kelly, Y. J., Nazroo, J. Y., McMunn, A., Boreham, R., & Marmot, M. (2001). Birthweight and behavioural problems in children: A modifiable effect? *International Journal of Epidemiology*, *30*(1), 88–94.

Kempermann, G., Kuhn, H. G., & Gage, F. H. (1997). More hippocampal neurons in adult mice living in an enriched environment. *Nature*, *386*(6624), 493.

Kim, J. J., & Diamond, D. M. (2002). The stressed hippocampus, synaptic plasticity and lost memories. *Nature Reviews. Neuroscience*, 3(6), 453.

Knudsen, E. I. (2004). Sensitive periods in the development of the brain and behavior. *Journal of Cognitive Neuroscience, 16*(8), 1412–1425. https://doi.org/10.1162/08989 29042304796.

Kobilo, T., Liu, Q.-R., Gandhi, K., Mughal, M., Shaham, Y., & van Praag, H. (2011). Running is the neurogenic and neurotrophic stimulus in environmental enrichment. *Learning & Memory, 18*(9), 605–609.

Koe, A., Ashokan, A., & Mitra, R. (2016). Short environmental enrichment in adulthood reverses anxiety and basolateral amygdala hypertrophy induced by maternal separation. *Translational Psychiatry, 6*(2). e729.

Koehnle, T. J., & Rinaman, L. (2010). Early experience alters limbic forebrain Fos responses to a stressful interoceptive stimulus in young adult rats. *Physiology & Behavior, 100*(2), 105–115. https://doi.org/10.1016/j.physbeh.2010.02.006.

Komitova, M., Zhao, L. R., Gidö, G., Johansson, B. B., & Eriksson, P. (2005). Postischemic exercise attenuates whereas enriched environment has certain enhancing effects on lesion-induced subventricular zone activation in the adult rat. *European Journal of Neuroscience, 21*(9), 2397–2405.

Korte, S. M., Koolhaas, J. M., Wingfield, J. C., & McEwen, B. S. (2005). The Darwinian concept of stress: Benefits of allostasis and costs of allostatic load and the trade-offs in health and disease. *Neuroscience & Biobehavioral Reviews, 29*(1), 3–38.

Krugers, H. J., Oomen, C. A., Gumbs, M., Li, M., Velzing, E. H., Joels, M., et al. (2012). Maternal deprivation and dendritic complexity in the basolateral amygdala. *Neuropharmacology, 62*(1), 534–537.

Laudenslager, M. L., Reite, M., & Harbeck, R. J. (1982). Suppressed immune response in infant monkeys associated with maternal separation. *Behavioral and Neural Biology, 36*(1), 40–48.

LeDoux, J. E. (2000). Emotion circuits in the brain. *Annual Review of Neuroscience, 23*, 155–184. https://doi.org/10.1146/annurev.neuro.23.1.155.

LeDoux, J. (2003). The emotional brain, fear, and the amygdala. *Cellular and Molecular Neurobiology, 23*(4–5), 727–738.

LeDoux, J. (2007). The amygdala. *Current Biology, 17*(20), R868–R874. https://doi.org/10.1016/j.cub.2007.08.005.

Lehmann, J., Pryce, C., Bettschen, D., & Feldon, J. (1999). The maternal separation paradigm and adult emotionality and cognition in male and female Wistar rats. *Pharmacology Biochemistry and Behavior, 64*(4), 705–715.

Levine, S. (1956). A further study of infantile handling and adult avoidance Learning1. *Journal of Personality, 25*(1), 70–80.

Lima, A. R., Mello, M. F., Andreoli, S. B., Fossaluza, V., de Araujo, C. M., Jackowski, A. P., et al. (2014). The impact of healthy parenting as a protective factor for posttraumatic stress disorder in adulthood: A case-control study. *PLoS One, 9*(1), e87117. https://doi.org/10.1371/journal.pone.0087117.

Lin, C.-H., Yeh, S.-H., Lin, C.-H., Lu, K.-T., Leu, T.-H., Chang, W.-C., et al. (2001). A role for the PI-3 kinase signaling pathway in fear conditioning and synaptic plasticity in the amygdala. *Neuron, 31*(5), 841–851.

Lindström, J. (1999). Early development and fitness in birds and mammals. *Trends in Ecology & Evolution, 14*(9), 343–348.

Liu, D., Diorio, J., Tannenbaum, B., Caldji, C., Francis, D., Freedman, A., et al. (1997). Maternal care, hippocampal glucocorticoid receptors, and hypothalamic-pituitary-adrenal responses to stress. *Science*, 277(5332), 1659.

Lupien, S. J., McEwen, B. S., Gunnar, M. R., & Heim, C. (2009). Effects of stress throughout the lifespan on the brain, behaviour and cognition. *Nature Reviews Neuroscience*, 10(6), 434.

Maccoby, E. E., & Masters, J. C. (1970). Attachment and dependency. *Carmichael's Manual of Child Psychology*, *2*, 73–157.
Magarin, A., & McEwen, B. (1995). Stress-induced atrophy of apical dendrites of hippocampal CA3c neurons: Involvement of glucocorticoid secretion and excitatory amino acid receptors. *Neuroscience*, *69*(1), 89–98.
Magariños, A. M. A., McEwen, B. S., Flügge, G., & Fuchs, E. (1996). Chronic psychosocial stress causes apical dendritic atrophy of hippocampal CA3 pyramidal neurons in subordinate tree shrews. *Journal of Neuroscience*, *16*(10), 3534–3540.
Mantis, I., Stack, D. M., Ng, L., Serbin, L. A., & Schwartzman, A. E. (2014). Mutual touch during mother-infant face-to-face still-face interactions: Influences of interaction period and infant birth status. *Infant Behavior & Development*, *37*(3), 258–267. https://doi.org/10.1016/j.infbeh.2014.04.005.
Maren, S., & Quirk, G. J. (2004). Neuronal signalling of fear memory. *Nature Reviews. Neuroscience*, 5(11), 844.
Mazuc, J., Chastel, O., & Sorci, G. (2003). No evidence for differential maternal allocation to offspring in the house sparrow (Passer domesticus). *Behavioral Ecology*, *14*(3), 340–346.
McEwen, B. S. (1998a). Protective and damaging effects of stress mediators. *New England Journal of Medicine*, *338*(3), 171–179.
McEwen, B. S. (1998b). Stress, adaptation, and disease: Allostasis and allostatic load. *Annals of the New York Academy of Sciences*, *840*(1), 33–44.
McEwen, B. S. (1999). Stress and hippocampal plasticity. *Annual Review of Neuroscience*, *22*(1), 105–122.
McEwen, B. S. (2000). The neurobiology of stress: From serendipity to clinical relevance. *Brain Research*, *886*(1), 172–189.
McEwen, B. S. (2003). Mood disorders and allostatic load. *Biological Psychiatry*, *54*(3), 200–207.
McEwen, B. S. (2007). Physiology and neurobiology of stress and adaptation: Central role of the brain. *Physiological Reviews*, *87*(3), 873–904.
McEwen, B. S. (2008). Central effects of stress hormones in health and disease: Understanding the protective and damaging effects of stress and stress mediators. *European Journal of Pharmacology*, *583*(2), 174–185.
McEwen, B. S., & Wingfield, J. C. (2003). The concept of allostasis in biology and biomedicine. *Hormones and Behavior*, *43*(1), 2–15.
McGaugh, J. L. (2004). The amygdala modulates the consolidation of memories of emotionally arousing experiences. *Annual Review of Neuroscience*, *27*, 1–28.
McNamara, J. M., & Houston, A. I. (1996). State-dependent life histories. *Nature*, *380*(6571), 215.
Mitra, R., Adamec, R., & Sapolsky, R. (2009). Resilience against predator stress and dendritic morphology of amygdala neurons. *Behavioural Brain Research*, *205*(2), 535–543.
Mitra, R., Jadhav, S., McEwen, B. S., Vyas, A., & Chattarji, S. (2005). Stress duration modulates the spatiotemporal patterns of spine formation in the basolateral amygdala. *Proceedings of the National Academy of Sciences of the United States of America*, *102*(26), 9371–9376.
Mitra, R., & Sapolsky, R. M. (2008). Acute corticosterone treatment is sufficient to induce anxiety and amygdaloid dendritic hypertrophy. *Proceedings of the National Academy of Sciences*, *105*(14), 5573–5578.
Mitra, R., & Sapolsky, R. M. (2012). Short-term enrichment makes male rats more attractive, more defensive and alters hypothalamic neurons. *PLoS One*, 7(5), e36092. https://doi.org/10.1371/journal.pone.0036092.
Moffett, M., Vicentic, A., Kozel, M., Plotsky, P., Francis, D., & Kuhar, M. (2007). Maternal separation alters drug intake patterns in adulthood in rats. *Biochemical Pharmacology*, *73*(3), 321–330.

Mohammed, A. H., Zhu, S. W., Darmopil, S., Hjerling-Leffler, J., Ernfors, P., Winblad, B., et al. (2002). Environmental enrichment and the brain. *Progress in Brain Research*, *138*, 109–133. https://doi.org/10.1016/S0079-6123(02)38074-9.

Monroy, E., Hernández-Torres, E., & Flores, G. (2010). Maternal separation disrupts dendritic morphology of neurons in prefrontal cortex, hippocampus, and nucleus accumbens in male rat offspring. *Journal of Chemical Neuroanatomy*, *40*(2), 93–101.

Moran, N. A. (1992). The evolutionary maintenance of alternative phenotypes. *The American Naturalist*, *139*(5), 971–989.

Moriceau, S., Roth, T. L., Okotoghaide, T., & Sullivan, R. M. (2004). Corticosterone controls the developmental emergence of fear and amygdala function to predator odors in infant rat pups. *International Journal of Developmental Neuroscience*, *22*(5–6), 415–422. https://doi.org/10.1016/j.ijdevneu.2004.05.011.

Moriceau, S., & Sullivan, R. M. (2006). Maternal presence serves as a switch between learning fear and attraction in infancy. *Nature Neuroscience*, *9*(8), 1004–1006. https://doi.org/10.1038/nn1733.

Mousseau, T. A., & Fox, C. W. (1998). *Maternal effects as adaptations*. Oxford University Press.

Munck, A., Guyre, P. M., & Holbrook, N. J. (1984). Physiological functions of glucocorticoids in stress and their relation to pharmacological actions. *Endocrine Reviews*, *5*(1), 25–44.

Murray, L., Stanley, C., Hooper, R., King, F., & Fiori-Cowley, A. (1996). The role of infant factors in postnatal depression and mother-infant interactions. *Developmental Medicine and Child Neurology*, *38*(2), 109–119.

Mustroph, M. L., Chen, S., Desai, S. C., Cay, E. B., DeYoung, E. K., & Rhodes, J. S. (2012). Aerobic exercise is the critical variable in an enriched environment that increases hippocampal neurogenesis and water maze learning in male C57BL/6J mice. *Neuroscience*, *219*, 62–71.

Muzik, M., Bocknek, E. L., Broderick, A., Richardson, P., Rosenblum, K. L., Thelen, K., et al. (2013). Mother–infant bonding impairment across the first 6 months postpartum: The primacy of psychopathology in women with childhood abuse and neglect histories. *Archives of Women's Mental Health*, *16*(1), 29–38.

Nestler, E. J., Barrot, M., DiLeone, R. J., Eisch, A. J., Gold, S. J., & Monteggia, L. M. (2002). Neurobiology of depression. *Neuron*, *34*(1), 13–25. https://doi.org/10.1016/S0896-6273(02)00653-0.

Nithianantharajah, J., & Hannan, A. J. (2006). Enriched environments, experience-dependent plasticity and disorders of the nervous system. *Nature Reviews. Neuroscience*, 7(9), 697.

Pacteau, C., Einon, D., & Sinden, J. (1989). Early rearing environment and dorsal hippocampal ibotenic acid lesions: Long-term influences on spatial learning and alternation in the rat. *Behavioural Brain Research*, *34*(1–2), 79–96.

Pearson, J. M., Watson, K. K., & Platt, M. L. (2014). Decision making: The neuroethological turn. *Neuron*, *82*(5), 950–965.

Pigliucci, M., Murren, C. J., & Schlichting, C. D. (2006). Phenotypic plasticity and evolution by genetic assimilation. *The Journal of Experimental Biology*, *209*(Pt. 12), 2362–2367. https://doi.org/10.1242/jeb.02070.

Plotsky, P. M., & Meaney, M. J. (1993). Early, postnatal experience alters hypothalamic corticotropin-releasing factor (CRF) mRNA, median eminence CRF content and stress-induced release in adult rats. *Brain Research. Molecular Brain Research*, *18*(3), 195–200.

Plotsky, P. M., Owens, M. J., & Nemeroff, C. B. (1998). Psychoneuroendocrinology of depression: Hypothalamic-pituitary-adrenal axis. *Psychiatric Clinics of North America*, *21*(2), 293–307.

Powell, J., Lewis, P. A., Roberts, N., Garcia-Finana, M., & Dunbar, R. I. (2012). Orbital prefrontal cortex volume predicts social network size: An imaging study of individual differences in humans. *Proceedings of the Biological Sciences*, *279*(1736), 2157–2162. https://doi.org/10.1098/rspb.2011.2574.

Radley, J. J., Arias, C. M., & Sawchenko, P. E. (2006). Regional differentiation of the medial prefrontal cortex in regulating adaptive responses to acute emotional stress. *Journal of Neuroscience*, *26*(50), 12967–12976.

Rainnie, D. G., Asprodini, E. K., & Shinnick-Gallagher, P. (1991a). Excitatory transmission in the basolateral amygdala. *Journal of Neurophysiology*, *66*(3), 986–998.

Rainnie, D. G., Asprodini, E. K., & Shinnick-Gallagher, P. (1991b). Inhibitory transmission in the basolateral amygdala. *Journal of Neurophysiology*, *66*(3), 999–1009.

Raison, C. L., & Miller, A. H. (2003). When not enough is too much: The role of insufficient glucocorticoid signaling in the pathophysiology of stress-related disorders. *American Journal of Psychiatry*, *160*(9), 1554–1565.

Reite, M., Kaemingk, K., & Boccia, M. L. (1989). Maternal separation in bonnet monkey infants: Altered attachment and social support. *Child Development*, *60*, 473–480.

Reite, M., Short, R., Kaufman, I., Stynes, A., & Pauley, J. (1978). Heart rate and body temperature in separated monkey infants. *Biological Psychiatry*.

Revest, J.-M., Di Blasi, F., Kitchener, P., Rougé-Pont, F., Desmedt, A., Turiault, M., et al. (2005). The MAPK pathway and Egr-1 mediate stress-related behavioral effects of glucocorticoids. *Nature Neuroscience*, *8*(5), 664.

Righetti-Veltema, M., Conne-Perréard, E., Bousquet, A., & Manzano, J. (2002). Postpartum depression and mother–infant relationship at 3 months old. *Journal of Affective Disorders*, *70*(3), 291–306.

Rivarola, M. A., & Renard, G. M. (2014). What we know about the long-term consequences of early maternal separation and neuroendocrine response to stress. *Revista de Farmacología de chile*, *17*, 17–25.

Rodrigues, S. M., LeDoux, J. E., & Sapolsky, R. M. (2009). The influence of stress hormones on fear circuitry. *Annual Review of Neuroscience*, *32*, 289–313.

Rogan, M. T., Stäubli, U. V., & LeDoux, J. E. (1998). Erratum: Fear conditioning induces associative long-term potentiation in the amygdala. *Nature*, *391*(6669), 818.

Romanski, L. M., & LeDoux, J. E. (1993). Information cascade from primary auditory cortex to the amygdala: Corticocortical and corticoamygdaloid projections of temporal cortex in the rat. *Cerebral Cortex*, *3*(6), 515–532.

Romeo, R. D., Bellani, R., Karatsoreos, I. N., Chhua, N., Vernov, M., Conrad, C. D., et al. (2006). Stress history and pubertal development interact to shape hypothalamic-pituitary-adrenal axis plasticity. *Endocrinology*, *147*(4), 1664–1674.

Roozendaal, B., McEwen, B. S., & Chattarji, S. (2009). Stress, memory and the amygdala. *Nature Reviews Neuroscience*, *10*(6), 423–433.

Russo, S. J., Murrough, J. W., Han, M.-H., Charney, D. S., & Nestler, E. J. (2012). Neurobiology of resilience. *Nature Neuroscience*, *15*(11), 1475–1484.

Rutstein, A. N., Gilbert, L., Slater, P. J., & Graves, J. A. (2004). Sex-specific patterns of yolk androgen allocation depend on maternal diet in the zebra finch. *Behavioral Ecology*, *16*(1), 62–69.

Rutter, M. (1972). *Maternal deprivation reassessed,* Penguin Books.

Rutter, M. (1985). Resilience in the face of adversity. Protective factors and resistance to psychiatric disorder. *The British Journal of Psychiatry*, *147*(6), 598.

Sabatini, M. J., Ebert, P., Lewis, D. A., Levitt, P., Cameron, J. L., & Mirnics, K. (2007). Amygdala gene expression correlates of social behavior in monkeys experiencing maternal separation. *Journal of Neuroscience*, *27*(12), 3295–3304.

Sale, A., Putignano, E., Cancedda, L., Landi, S., Cirulli, F., Berardi, N., et al. (2004). Enriched environment and acceleration of visual system development. *Neuropharmacology, 47*(5), 649–660.

Sanders, B. J., & Anticevic, A. (2007). Maternal separation enhances neuronal activation and cardiovascular responses to acute stress in borderline hypertensive rats. *Behavioural Brain Research, 183*(1), 25–30. https://doi.org/10.1016/j.bbr.2007.05.020.

Sapolsky, R. M., Krey, L. C., & McEwen, B. S. (1984). Glucocorticoid-sensitive hippocampal neurons are involved in terminating the adrenocortical stress response. *Proceedings of the National Academy of Sciences, 81*(19), 6174–6177.

Sapolsky, R. M., Krey, L. C., & McEwen, B. S. (1986). The neuroendocrinology of stress and aging: The glucocorticoid cascade hypothesis. *Endocrine Reviews*, 7(3), 284–301.

Schore, A. N. (2001). Effects of a secure attachment relationship on right brain development, affect regulation, and infant mental health. *Infant Mental Health Journal, 22*(1–2), 7–66. Official Publication of The World Association for Infant Mental Health.

Schulz, L. C. (2010). The Dutch hunger Winter and the developmental origins of health and disease. *Proceedings of the National Academy of Sciences, 107*(39), 16757–16758.

Seay, B., Alexander, B. K., & Harlow, H. F. (1964). Maternal behavior of socially deprived rhesus monkeys. *The Journal of Abnormal and Social Psychology, 69*(4), 345.

Selye, H. (1936). A syndrome produced by diverse nocuous agents. *Nature, 138*(3479), 32.

Shaffer, D. D. R., & Kipp, K. (2002). *Developmental psychology: Childhood & adolescence.* Cengage learning.

Smith, S. M., & Vale, W. W. (2006). The role of the hypothalamic-pituitary-adrenal axis in neuroendocrine responses to stress. *Dialogues in Clinical Neuroscience, 8*(4), 383.

Sofroniew, M. V., Howe, C. L., & Mobley, W. C. (2001). Nerve growth factor signaling, neuroprotection, and neural repair. *Annual Review of Neuroscience, 24*, 1217–1281. https://doi.org/10.1146/annurev.neuro.24.1.1217.

Spencer, S. J., Buller, K. M., & Day, T. A. (2005). Medial prefrontal cortex control of the paraventricular hypothalamic nucleus response to psychological stress: Possible role of the bed nucleus of the stria terminalis. *Journal of Comparative Neurology, 481*(4), 363–376.

Sroufe, L. A. (1997). *Emotional development: The organization of emotional life in the early years.* Cambridge University Press.

Suderman, M., McGowan, P. O., Sasaki, A., Huang, T. C. T., Hallett, M. T., Meaney, M. J., et al. (2012). Conserved epigenetic sensitivity to early life experience in the rat and human hippocampus. *Proceedings of the National Academy of Sciences*, 109(Suppl. 2), 17266–17272.

Sultan, S. E. (2003). Commentary: The promise of ecological developmental biology. *Journal of Experimental Zoology Part B: Molecular and Developmental Evolution, 296*(1), 1–7.

Sultan, S. E., & Spencer, H. G. (2002). Metapopulation structure favors plasticity over local adaptation. *The American Naturalist, 160*(2), 271–283.

Suomi, S. J. (2011). Risk, resilience, and gene-environment interplay in primates. *Journal of Canadian Academy of Child and Adolescent Psychiatry, 20*(4), 289–297.

Tau, G. Z., & Peterson, B. S. (2010). Normal development of brain circuits. *Neuropsychopharmacology, 35*(1), 147–168. https://doi.org/10.1038/npp.2009.115.

Tierney, A. L., & Nelson, C. A., III. (2009). Brain development and the role of experience in the early years. *Zero to three, 30*(2), 9.

Tryon, R. C. (1940). *Genetic differences in mazelearning ability in rats.* Yearbook of the National Society for the Study of Education.

Tufto, J. (2000). The evolution of plasticity and nonplastic spatial and temporal adaptations in the presence of imperfect environmental cues. *The American Naturalist, 156*(2), 121–130.

Vallee, M., Mayo, W., Dellu, F., Le Moal, M., Simon, H., & Maccari, S. (1997). Prenatal stress induces high anxiety and postnatal handling induces low anxiety in adult offspring: Correlation with stress-induced corticosterone secretion. *Journal of Neuroscience, 17*(7), 2626–2636.

van Praag, H., Christie, B. R., Sejnowski, T. J., & Gage, F. H. (1999). Running enhances neurogenesis, learning, and long-term potentiation in mice. *Proceedings of the National Academy of Sciences of the United States of America, 96*(23), 13427–13431.

van Praag, H., Kempermann, G., & Gage, F. H. (2000). Neural consequences of envi-romental enrichment. *Nature Reviews. Neuroscience, 1*(3), 191–198. https://doi.org/10.1038/35044558.

Vouloumanos, A., & Waxman, S. R. (2014). Listen up! Speech is for thinking during infancy. *Trends in Cognitive Sciences, 18*(12), 642–646. https://doi.org/10.1016/j.tics.2014.10.001.

Vyas, A., & Chattarji, S. (2004). Modulation of different states of anxiety-like behavior by chronic stress. *Behavioral Neuroscience, 118*(6), 1450.

Vyas, A., Mitra, R., Rao, B. S., & Chattarji, S. (2002). Chronic stress induces contrasting patterns of dendritic remodeling in hippocampal and amygdaloid neurons. *Journal of Neuroscience, 22*(15), 6810–6818.

Weaver, I. C., Cervoni, N., Champagne, F. A., D'Alessio, A. C., Sharma, S., Seckl, J. R., et al. (2004). Epigenetic programming by maternal behavior. *Nature Neuroscience*, 7(8), 847–854.

Weaver, I. C., Meaney, M. J., & Szyf, M. (2006). Maternal care effects on the hippocampal transcriptome and anxiety-mediated behaviors in the offspring that are reversible in adulthood. *Proceedings of the National Academy of Sciences of the United States of America, 103*(9), 3480–3485.

Weidenfeld, J., & Ovadia, H. (2017). The role of the amygdala in regulating the hypothalamic-pituitary-adrenal Axis. In *The amygdala-where emotions shape perception, learning and memories*: InTech.

William Huck, U., Labov, J. B., & Lisk, R. D. (1987). Food-restricting first generation juvenile female hamsters (*Mesocricetus auratus*) affects sex ratio and growth of third generation offspring. *Biology of Reproduction, 37*(3), 612–617.

Yang, J., Hou, C., Ma, N., Liu, J., Zhang, Y., Zhou, J., et al. (2007). Enriched environment treatment restores impaired hippocampal synaptic plasticity and cognitive deficits induced by prenatal chronic stress. *Neurobiology of Learning and Memory, 87*(2), 257–263.

Young, D., Lawlor, P. A., Leone, P., Dragunow, M., & During, M. J. (1999). Environmental enrichment inhibits spontaneous apoptosis, prevents seizures and is neuroprotective. *Nature Medicine, 5*(4), 448–453. https://doi.org/10.1038/7449.

Zalewski, M., Lengua, L. J., Kiff, C. J., & Fisher, P. A. (2012). Understanding the relation of low income to HPA-axis functioning in preschool children: Cumulative family risk and parenting as pathways to disruptions in cortisol. *Child Psychiatry & Human Development, 43*(6), 924–942.

CHAPTER EIGHT

Stress, the cortisol awakening response and cognitive function

Robin Law*, Angela Clow
Psychology, School of Social Sciences, University of Westminster, London, England
*Corresponding author: e-mail address: r.law2@westminster.ac.uk

Contents

Abstract

There is evidence that stress-induced disruption of the circadian rhythm of cortisol secretion, has negative consequences for brain health. The cortisol awakening response (CAR) is the most prominent and dynamic aspect of this rhythm. It has complex regulatory mechanisms making it distinct from the rest of the cortisol circadian rhythm, and is frequently investigated as a biomarker of stress and potential intermediary between stress and impaired brain function. Despite this, the precise function of the CAR within the healthy cortisol circadian rhythm remains poorly understood. Cortisol is a powerful hormone known to influence cognition in multiple and complex ways. Studies of the CAR and cognitive function have used varied methodological approaches which have produced similarly varied findings. The present review considers the accumulating evidence linking stress, attenuation of the CAR and reduced cognitive function, and seeks to contextualize the many findings to study populations, cognitive measures, and CAR methodologies employed. Associations between the CAR and both memory and executive functions are discussed in relation to its potential role as a neuroendocrine time of day signal that synchronizes peripheral clocks throughout the brain to enable optimum function, and recommendations for future research are provided.

International Review of Neurobiology, Volume 150
ISSN 0074-7742
https://doi.org/10.1016/bs.irn.2020.01.001

1. Stress, cortisol and cognition

Stress is known to have marked and varied effects on cognitive function (McEwen, 2012; McEwen & Gianaros, 2011; McEwen & Sapolsky, 1995) and these effects are largely attributed to the actions of the hormone cortisol on brain function. Cortisol is a powerful steroid hormone released from the adrenal cortex into the circulation following activation of the hypothalamic-pituitary-adrenal (HPA) axis by stress and the hypothalamic suprachiasmatic nucleus (SCN).

The effects of cortisol on cognitive functions in humans are seen in response to both acute and chronic elevated secretion (see Buchanan & Lovallo, 2001; Lupien et al., 1994; Mizoguchi, Ishige, Takeda, Aburada, & Tabira, 2004; Vedhara, Hyde, Gilchrist, Tytherleigh, & Plummer, 2000). These can be either enhancing or impairing effects depending upon the specific cognitive function, and the timing and quantity of cortisol exposure (Domes, Rothfischer, Reichwald, & Hautzinger, 2005; Kirschbaum, Wolf, May, Wippich, & Hellhammer, 1996; Schilling et al., 2013; Yehuda, Harvey, Buchsbaum, Tischler, & Schmeidler, 2007). The extent to which cortisol influences different functions is, in part, a reflection of the number and type of cortisol receptors found in the associated brain regions. Two separate intracellular receptors are expressed within the body that bind with cortisol: Glucocorticoid receptors (GR) and mineralocorticoid receptors (MR). GR are found in most tissues, whereas MR are found in the brain, kidney, colon, heart and sweat glands (Joëls, Karst, DeRijk, & de Kloet, 2008). The MR shows a similar intrinsic affinity for aldosterone, cortisol and corticosterone, while the GR has reduced intrinsic affinity for glucocorticoids in comparison (Gomez-Sanchez, 2010). Within the cell, MR and GR activation interacts in either a complementary or opposing fashion depending on the cell type, and this interaction is crucial in determining cell function (Gomez-Sanchez, 2014).

An important moderator of acute cortisol and cognition associations is the timing of the cognitive measure relative to cortisol secretion. Cortisol exerts effects on neuronal activity via both rapid-acting, non-genomic (independent of alterations in gene expression) and slow-acting, genomic (mediated by alterations in gene expression) effects (Joëls, Fernandez, & Roozendaal, 2011). This is an adaptive feature as, in the context of a stress-response, it facilitates both immediate enhancement of cognitive functions required to respond to the immediate challenge, and slower effects to influence cognition when different demands may be faced. For example, a meta-analysis of acute cortisol administration and executive functions conducted by Shields, Bonner, and Moons (2015) showed that rapid, non-genomic effects of cortisol enhanced inhibition, while slower, genomic effects impaired inhibition. As such, the time-dependent nature of this relationship may contribute substantially to the variability in results of studies of cortisol effects on cognition.

The hippocampus, pre-frontal cortex, and amygdala all show high levels of both MR and GR receptors compared to other brain regions (De Kloet et al., 2000; Herbert et al., 2006). The effects of cortisol on the functioning of these regions is determined by the balance of the MR binding relative to GR binding. For example, basal levels of cortisol binding to MR receptors in the hippocampus (HC) in the temporal lobe has been shown to enhance excitability, and elevated levels of cortisol binding to the GR during stress responding have the opposite effect (De Kloet, Oitzl, & Joëls, 1999; Herbert et al., 2006). Longer-term regulation of brain structure and function by prolonged exposure to cortisol is also achieved by alteration of the density of neurons through apoptosis and inhibited neurogenesis (Pittenger & Duman, 2008), and by influencing dendritic complexity (McEwen, 2008). Glucocorticoids play an important role in mediating the remodeling of neurons in the hippocampus and frontal cortex during repeated stress exposure, likely due to regulatory effects on glutamate (McEwen, 1999, 2008; Vyas, Mitra, Rao, & Chattarji, 2002). Such actions make cortisol, along with other mediators such as serotonin, GABA, and excitatory amino acids, an important modulator of brain plasticity and cognitive functions (McEwen, 1999, 2008).

2. The cortisol awakening response

Cortisol secretion in healthy humans shows a marked circadian rhythm, characterized by a gradual increase during the late part of sleep to peak after morning awakening, then declining throughout the day to the nadir in the late evening and early part of sleep (Linkowski et al., 1985; Weitzman et al., 1971). The cortisol awakening response (CAR) is the dynamic increase in cortisol concentrations within the first hour after awakening from night-time sleep, peaking around 30- to 40-min post-awakening (Clow, Thorn, Evans, & Hucklebridge, 2004; Pruessner et al., 1997). The CAR is seen from early infancy to older adulthood (Fries, Dettenborn, & Kirschbaum, 2009; Stalder et al., 2013), and initial studies indicated that it is observed in around 77% of the healthy adult population (Wüst et al., 2000), though this figure appears far higher when appropriate study methodology is employed (e.g., Oskis, Loveday, Hucklebridge, Thorn, & Clow, 2009; Smyth, Thorn, Hucklebridge, Clow, & Evans, 2016). Within the circadian rhythm of cortisol secretion the CAR is a relatively discrete aspect, since it is initiated by morning awakening and is superimposed upon the circadian rhythm (Clow et al., 2004; Clow, Hucklebridge, Stalder, Evans, & Thorn, 2010; Wilhelm, Born, Kudielka, Schlotz, & Wüst, 2007). This has been demonstrated by investigation of measures of pre- and post-awakening levels of cortisol and ACTH in blood and saliva showing that steep increases are seen in secretion of both hormones in response to awakening, and that these are distinct from baseline circadian secretion (Wilhelm et al., 2007). The CAR is also relatively independent of cortisol secretion across the rest of the day (Edwards, Clow, Evans, & Hucklebridge, 2001; Maina, Palmas, Bovenzi, & Filon, 2009; Schmidt-Reinwald et al., 1999). This is thought to be because regulation of the CAR involves additional direct modulation of the adrenal cortex by the SCN (via the pathway described previously), which enhances adrenal sensitivity in the immediate post-awakening period (Clow, Hucklebridge, Stalder, et al., 2010).

The CAR is a dynamic post-awakening increase in cortisol secretion and therefore its measurement requires two components: cortisol levels immediately post-awakening, and the post-awakening rise (Stalder et al., 2016). For example, it can be calculated as the 'area under the curve with respect to increase' (AUCi), 'mean increase' (MnInc), and simple delta measures (i.e., peak value minus that upon awakening). But measures of total cortisol levels in the post-awakening period, such as the 'area under the curve with respect to ground' (AUCg), should not be considered as CAR measures as they lack sensitivity to this dynamic increase (for detailed description of CAR measures, see Stalder et al., 2016).

The CAR is considered a key link between mind and body due to its sensitivity to psychosocial factors such as negative affect and anticipation of workload in the day ahead (Clow, Hucklebridge, Stalder, et al., 2010; Fries et al., 2009). Since first identification of the CAR in humans by Pruessner et al. (1997), it has been researched extensively in the context of understanding its regulation and its implications as a biomarker of health status (Clow et al., 2004; Clow, Hucklebridge, & Thorn, 2010; Fries et al., 2009). From this research it has been established that the magnitude of the CAR is associated with a range of physiological and psychological factors, at both an inter- and intra-individual level. Although it was initially thought that the CAR was heightened in chronic stress (Chida & Steptoe, 2009) more reliable methodologies consistently report attenuation with chronic stress (e.g., Duan et al., 2013; Juster et al., 2011; Lovell, Moss, & Wetherell, 2011; O'Connor et al., 2009; Saban, Mathews, Bryant, O'Brien, & Janusek, 2012). This observation is consistent with reports of a flattened cortisol diurnal decline in a range of negative physical and psychological health outcomes (Adam et al., 2017). In other words, chronic stress leads to a reduced dynamic range in the circadian rhythm of cortisol secretion, including the CAR.

Regarding physiological factors, it has been established that CAR magnitude shows a negative association with age (Heaney, Phillips, & Carroll, 2010; Knoops, van der Graaf, Willem, & Geerlings, 2010) and varies by sex (Oskis et al., 2009; Wright & Steptoe, 2005; Wüst, Wolf, et al., 2000). Notably sex differences have been observed in the timing of the CAR peak, with average peak concentrations seen at 30-min post-awakening in males and at 45-min post-awakening in females (Oskis et al., 2009). Females also typically show a delayed decline in cortisol post-CAR peak resulting in greater overall cortisol secretion in the post-awakening hour (Wüst, Wolf, et al., 2000). Genetic factors also appear to play an important role in determining interindividual differences in the CAR; Wüst, Federenko, Hellhammer, and Kirschbaum (2000) reported results from a twin study revealing a heritiability index estimate of 0.40 for dynamic increase in cortisol levels post-awakening. While, with regard to psychosocial factors, it has been shown that in healthy participants a greater CAR magnitude is seen both in response to negative prior-day experiences and increased anticipations of challenge in the coming day (Adam, Hawkley, Kudielka, & Cacioppo, 2006; Clow, Hucklebridge, & Thorn, 2010; Doane & Adam, 2010; Stalder, Evans, Hucklebridge, & Clow, 2010a, 2010b).

CAR research to date has primarily focused on its application as a biomarker in association with a range of psychosocial and physical variables including health outcomes (Chida & Steptoe, 2009). In addition to attenuation in chronic stress cross-sectional studies have reported blunted CAR profiles in cardiovascular, autoimmune, allergic, and psychiatric disorders (Clow et al., 2004; Fries et al., 2009; Wüst, Wolf, et al., 2000). As for stress research results of such studies in clinical populations have often been inconsistent, perhaps due to participant inaccuracy in CAR sampling (i.e., delays in collecting the first waking sample result in erroneous CAR estimates), differences in experimental design, and differences in sample demographics such as age, gender and genotype (Clow, Hucklebridge, & Thorn, 2010; Smyth et al., 2016; Stalder et al., 2016). The CAR is also typically used as a trait (inter-individual) measure, despite being subject to substantial state (within-individual) variation, which presents another potentially challenge for interpretation of results from between-subjects studies (for review, see Law, Hucklebridge, Thorn, Evans, & Clow, 2013).

Most CAR studies rely upon participants' self-collection of saliva samples within the domestic setting, and validity of results therefore depend upon accuracy of sampling. However, inaccurate reporting of awakening and/or sampling times is relatively common, and failure to effectively monitor such inaccuracy has been recognized as a serious confound in this area (for review, see Stalder et al., 2016). This is particularly the case if sampling delay exceeds the beginning of the dynamic increase in cortisol section, which occurs ~8-min post-awakening (Clow et al., 2004; Smyth et al., 2016). The recent consensus guidelines for CAR research (Stalder et al., 2016) propose objective assessment of sampling accuracy using electronic monitoring of both awakening and sampling times. While detailed assessment of the effects of sampling inaccuracy is not the focus of the present paper, it is important to consider that many of the studies of stress, the CAR, and cognition were conducted prior to there being detailed knowledge of these methodological issues. Given the importance of this potential confound, the present review accounts for whether electronic monitoring of awakening was included in the studies described (see Table 1).

Another factor associated with within-subject, day-to-day variation in CAR magnitude is pre- and post-awakening light exposure (e.g., Petrowski, Schmalbach, Niedling, & Stalder, 2019; Scheer & Buijs, 1999; Thorn, Hucklebridge, Esgate, Evans, & Clow, 2004). In two recent and well-controlled sleep laboratory experiments in males, Petrowski et al. (2019) explored the differential effects of immediate post-awakening light exposure on the CAR and found consistent stimulatory effects of both blue and green wavelength light. The positive association between CAR magnitude and light exposure is indicative of the modulatory influence of the SCN on the CAR, and may have implications for a role of the CAR within healthy function as a potential time-of-day marker for peripheral clocks of the circadian system, under the influence of the SCN (Clow, Hucklebridge, Stalder, et al., 2010; Karatsoreos, 2014; Law et al., 2013; Menet & Rosbash, 2011).

The SCN regulates the CAR via two separate pathways; the previously described input to the Paraventricular Nucleus PVN and the HPA axis

Table 1 Summary of CAR-cognition studies.

Author	Sample	CAR Measure	Cognitive Test	Findings	Notes
Memory only					
Pruessner, Pruessner, Hellhammer, Pike, and Lupien (2007)	13 healthy males (age 19–32, mean = 23.85)	AUC of 0–30–60 (4 days, at weekly intervals)	DM (cued recall) before and after a stressor	No significant relationship	Unclear if AUCg or AUCi used. No electronic monitoring of sampling accuracy
Bäumler, Voigt, et al. (2014)	97 children (ages 3–6)	0–30 delta (3 days)	Prospective memory (ball sorting)	CAR (+) prospective memory	Electronically monitored sampling accuracy
Bäumler, Kliegel, et al. (2014)	35 children (ages 3–6)	0–30 delta (2 days)	Naturalistic prospective memory (remembering to perform an action)	CAR (+) when prospective memory performance required	Electronically monitored sampling accuracy
Almela, van der Meij, Hidalgo, Villada, and Salvador (2012)	88 healthy middle aged adults (ages 55–77)	AUCi and AUCg (2 days)	DM and WM	AUCi, AUCg (−) DM *in men only*, CAR (+) WM	Those with negative CAR also showed poorer DM. Did not electronically monitor sampling accuracy
Rimmele, Meier, Lange, and Born (2010)	16 healthy males (mean age 22.3, SD 3.89)	'Morning cortisol' 7:00–8:30 am at 15 min intervals (1 day)	DM (free recall)	Inverted U-shaped relationship: extreme high or low CAR (−) DM	Experiment conducted within a laboratory

Continued

Table 1 Summary of CAR-cognition studies.—cont'd

Author	Sample	CAR Measure	Cognitive Test	Findings	Notes
Moriarty et al. (2014)	19 males age 30–60 (mean 40.6, SD 5.8)	AUCg and AUCi (1 day)	Spatial working memory (Newcastle 2D SM test) and Attention networks (broadly EF)	U-shaped relationship, extreme high or low AUCg (−) SWM	No relationship with AUCi (CAR). Did not electronically monitor sampling accuracy
Hinkelmann et al. (2013)	41 depressed patients, 41 controls (ages 18–70)	AUCg and AUCi (2 days)	Verbal learning and visuospatial memory	In depressives, AUCi and AUCg (−) both mem scores. In controls, CAR (+) verbal mem	Broad age range. Did not electronically monitor sampling accuracy
Hodyl et al. (2016)	39 healthy adults (mean age 22, SD 4)	AUCi (2 days)	Serial sequence reaction time task	CAR (−) learning and speed of performance	Electronically monitored awakening, but not sampling times
Executive function only					
Evans, Hucklebridge, Loveday, and Clow (2012)	50 older adults (ages 60–91)	MnInc, and average peak time (2 days)	Trail making	Early peak + CAR (+) EF	Electronically monitored sampling accuracy
Law, Evans, Thorn, Hucklebridge, and Clow (2015)	Case study of healthy young adult male	0–30 min rise (50 days)	Attention switching task	Within-subject CAR (+) attention switching performance	Single participant case study. Electronically monitored sampling accuracy

Zhang et al. (2015)	63 healthy young males	AUCi (sum of 2 days)	Go/no go task	CAR (+) error-related negativity latency and post-error miss rate	No electronic monitoring of sampling accuracy
Dierolf et al. (2016)	40 healthy males, ages 18–28 (mean age 23, SD 2.9)	Mean AUCg (3 days)	Emotion vs. gender task switching	AUCg (−) task switching performance	Did not examine dynamic (CAR) measure. No monitoring of sampling accuracy
Butler, Klaus, Edwards, and Pennington (2017)	109 healthy males, ages 21–63 (mean age 34.2, SD 10.6)	0–30 delta (2 days)	Accuracy of planning/problem solving (Tower of London-type task and set shifting)	CAR (−) problem solving. No significant association with task switching or WM	12% of awakening times were electronically monitored
Shi et al. (2018)	47 healthy males, ages 18–26 (mean age 22.4, SD 1.9)	Peak concentration minus that on awakening (1 day)	Go/no go task	CAR (+) response inhibition	Measured CAR on one day only. Electronically monitored sampling accuracy
Law et al. (forthcoming)	55 healthy young adults (mean age 20.2, SD 3.0)	MnInc (2 days)	Attention switching task	CAR (+) attention switching performance	Electronically monitored sampling accuracy
Various cognitive measures					
Maldonado et al. (2008)	116 children, ages 9–12, defined as either low or high stress	0–30 delta (1 day)	Cognitive drug research assessment system	CAR (−) speed of memory	Sampling supervised by participants' parents, without electronic monitoring of accuracy

Continued

Table 1 Summary of CAR-cognition studies.—cont'd

Author	Sample	CAR Measure	Cognitive Test	Findings	Notes
Stawski et al. (2011)	1500 midlife adults (mean age 57, SD 12)	0–30 min 'Morning rise' (4 days)	Overall measure inc. word recall, working mem, reasoning and proc. speed	No relationship with dynamic measure, but overall cortisol levels (+) overall cog	No age differences in observed relationship. Electronically monitored approximately 25% of sampling, but not awakening times
Evans et al. (2011)	50 older adults, ages 60–91 (mean 74)	Mean 0–45, and 0–30 delta (2 days)	Overall cog perf (OCP)	CAR (+) overall cog	Electronic monitoring of sampling accuracy
Franz et al. (2011)	795 male twins ages 51–60 (mean 55.9, SD 2.6)	0–30 delta (3 days, non-consecutive)	Range including spatial abilities, short- and long-term mem, EF, verbal fluency, reasoning, and processing speed	CAR (−) visual-spatial memory	Reported that the effect was entirely driven by overall diurnal cortisol. Participants received electronic sampling time reminders, but no electronic monitoring of accuracy
Aas et al. (2011)	30 Patients with first episode psychosis, 26 controls (ages 18–65)	AUCi (1 day)	Range including OCP, verbal mem, spatial abilities, processing speed, EF WM	In patients, CAR (+) verbal memory and processing speed	No relationship between CAR and cog in healthy controls. Did not monitor sampling accuracy

Cullen et al. (2014)	Children at risk for psychosis, and healthy control group (age range 11–14)	AUCi (2 days)	Broad range of memory and EF tests	In at risk groups, CAR (+) letter fluency and verbal memory	No relationship in healthy controls. Did not monitor sampling accuracy
Oosterholt, Maes, Van der Linden, Verbraak, and Kompier (2016)	85 adults (31 with clinical burnout, 27 non-clinical burnout, 27 healthy controls)	AUCg and 0–30 delta (2 days)	EF (prepotent response inhibition, irrelevant information inhibition, and task switching), verbal mem, cognitive failures (self-report)	Burnout patients (−) CAR, also worse updating (measured by 2-back task), and more cognitive failures	Did not directly compare cortisol and cognition measures, and did not monitor CAR sampling accuracy
Ennis, Moffat, and Hertzog (2016)	56 healthy adults, ages 23–79 (mean = 53, SD = 16.9)	0–30 delta divided by 0–30 sample-time delta (10 days)	Episodic mem, Working mem, processing speed. All measured 8–38 months after CAR	CAR (+) episodic memory, but no association with working mem or processing speed	Did not electronically monitor sampling accuracy
Hidalgo, Almela, Pulopulos, and Salvador (2016)	64 healthy adults, ages 57–76 (mean = 64.7, SD = 4.1)	AUCi (2 days)	Logical mem, verbal paired associates, family pictures test, letter-number sequencing, digit span, spatial span	CAR (−) immediate verbal and visual recall, but no association with any other measures	No significant effects with appropriate controls applied. Electronically monitored sampling but not awakening time

Continued

Table 1 Summary of CAR-cognition studies.—cont'd

Author	Sample	CAR Measure	Cognitive Test	Findings	Notes
Labad et al. (2016)	60 patients with early psychosis, ages 18–35	AUCi (1 day)	Range including OCP, WM, verbal and visual mem, reasoning and problem solving, social cognition	CAR magnitude (−) OCP, and flatness of CAR (−) spatial memory	Did not monitor sampling accuracy, and measured CAR on 1 day only (different day to cognitive test)
Salvat-Pujol et al. (2017)	97 medicated patients with MDD (mean age = 59.8, SD = 11.7), 97 healthy controls (mean age = 56.6, SD = 11.9)	AUCi (1 day)	Range of tests, including Hopkins verbal learning test, trail making, Rey complex figure test, Stroop, WM span	CAR (+) trail making, processing speed and Rey complex figure for MDD in remission	Did not monitor sampling accuracy, and measured CAR on 1 day only. No relationship with any measure in overall sample

AUCi, Area under the curve with respect to increase; AUCg, Area under the curve with respect to ground; OCP, Overall cognitive performance; WM, Working memory; DM, Declarative memory; EF, Executive function; Mem, Memory; MDD, Major depressive disorder; SD, Standard deviation; (+), Associated increase; (−), Associated decrease.

cascade (corticotropin-releasing hormone and adrenocorticotropic hormone), and also a direct autonomic input to the adrenal cortex via the splanchnic nerve (Clow, Hucklebridge, & Thorn, 2010). This dual-pathway SCN input plays an important role in determining CAR magnitude by a combination of pre- and post-awakening influences. While the HPA axis encourages cortisol secretion by increasing circulating ACTH in anticipation of awakening, the direct SCN-adrenal pathway is implicated in the fine-tuning of the CAR (Buijs, Eden, Goncharuk, & Kalsbeek, 2003). It achieves this by encouraging reduced adrenal sensitivity to ACTH in the immediate pre-awakening period, and then reversing this upon awakening such that adrenal sensitivity to ACTH is increased in the immediate post-awakening period (Fehm, Klein, Holl, & Voigt, 1984; Buijs, Wortel, Van Heerikhuize, & Kalsbeek, 1997; Buijs et al., 2003; Bornstein, Engeland, Ehrhart-Bornstein, & Herman, 2008). Direct innervation from SCN to adrenal gland is also implicated in CAR regulation due to the increased adrenal sensitivity to light in this period, as in rats this has been demonstrated to be both independent of circulating ACTH and dependent upon SCN integrity (Buijs et al., 1999).

Despite considerable research interest in the CAR for over 20 years, a precise role for the CAR within the healthy function has yet to be elucidated. The most prominent theory for the function of the CAR is that it provides an allostatic 'boost' upon awakening, to help the individual deal with the anticipated demands of the coming day (Adam et al., 2006; Clow, Hucklebridge, & Thorn, 2010; Clow, Hucklebridge, Stalder, et al., 2010; Clow et al., 2014; Fries et al., 2009; Law et al., 2013). It is thought that a morning 'boost' provided by the CAR may involve modulation of cognition (see Clow, Hucklebridge, Stalder, et al., 2010; Fries et al., 2009). This is in part due to correlational evidence of blunted or absent CARs in various disorders involving cognitive impairment. However, a causal relationship is feasible due to the many associations between cortisol secretion and cognition that are described within the present review. If true, this would mean that the CAR may not just be a biomarker of stress but play a role in the pathway from chronic stress to cognitive dysfunction.

3. The suprachiasmatic nucleus

The circadian rhythm of cortisol secretion plays a major role in synchronizing biological functions to appropriate times of day around the sleep-wake and the light-dark cycles (Oster et al., 2016). The suprachiasmatic nucleus (SCN) of the hypothalamus acts as the circadian pacemaker in humans and synchronizes circadian cortisol secretion to exogenous zeitgebers. Importantly, the SCN itself does not express GR and therefore acts as a cortisol output signaller only (Herbert et al., 2006). The SCN receives light information from the eye, though this photic input requires neither rods nor cones (the only previously known retinal photoreceptors), but instead a novel class of intrinsically photoreceptive melanopsin expressing retinal ganglion cells (Berson, 2007; Berson, Dunn, & Takao, 2002). These photosensitive cells project directly to the SCN via the retinohypothalamic tract (Berson, 2007; Herbert et al., 2006).

Exposure to light stimulates a cascade of molecular events in the SCN involving clock gene expression, which facilitates the entrainment of endogenous circadian rhythms to the light-dark cycle and allows for adaptation to seasonal changes in day length (Bunney and Bunney, 2000). Animal studies have shown that if the SCN is removed, the duration of sleep and wake remain similar, but that these behaviors no longer show a regular cycle (Buijs et al., 2003). As such the SCN does not directly control these behaviors, but simply plays a role in synchronizing them with the external environment.

Regarding cortisol secretion, the SCN is connected to the PVN of the hypothalamus both directly by axonal projections from SCN neurons, and indirectly via the SCN output pathway to the dorsomedial hypothalamus (Herbert et al., 2006). The SCN also exerts direct autonomic influence over cortisol secretion via the splanchnic nerve (for review see Clow et al., 2010; Clow, Hucklebridge, Stalder, Evans, Thorn, 2010). This fast-acting, non-HPA dependent pathway is thought to be particularly important with regards to modulation of circadian rhythms as it mediates adrenal sensitivity to ACTH according to the time of day (Bornstein et al., 2008; Buijs et al., 1997; Buijs et al., 2003; Clow, Hucklebridge, Stalder, et al., 2010). Thus, the SCN's light-responsive mediation of both the HPA-axis and non-HPA pathways helps to ensure fine-tuning of cortisol secretion that is appropriately synchronized to day-night cycles; stimulating secretion during the daytime and inhibiting it during the night (Benarroch, 2011).

While the SCN acts as the central circadian pacemaker, there are 'peripheral clocks' in organs throughout the body, including the brain. These peripheral clocks show 'free-running' circadian rhythms in isolation but are synchronized by indirect signaling from the SCN. It has been established that the SCN entrains endocrine and behavioral rhythms (e.g., glucocorticoids, melatonin; sleep-wake, and body temperature) via a range of signaling methods, which include direct influence by neuronal input and indirect influence via regulation of paracrine signals. In turn, these rhythms act to synchronize peripheral clocks, including those in the brain (Antle & Silver, 2005; Menet & Rosbash, 2011). As the authors have noted previously (Clow, Hucklebridge, Stalder, et al., 2010; Law et al., 2013) this suggests a role for the circadian cortisol rhythm, including the CAR, in entrainment of the peripheral circadian clocks in the body and brain. This hypothesis is supported by evidence that human peripheral clocks are entrained by glucocorticoids (Cuesta, Cermakian, & Boivin, 2015), and that the circadian rhythm of circulating glucocorticoids works to synchronize circadian rhythms of peripheral clocks with the SCN master clock (Oster et al., 2016).

4. The circadian system and psychopathology

The importance of the circadian rhythm of cortisol secretion is highlighted by evidence that disruption to this rhythm is associated with a broad range of both psychological and somatic conditions, including cognitive deficits (Lupien, McEwen, Gunnar, & Heim, 2009), post-traumatic stress disorder (Yehuda, Teicher, Trestman, Levengood, & Siever, 1996), chronic stress (Miller, Chen, & Zhou, 2007), burnout (Pruessner, Hellhammer, & Kirschbaum, 1999), chronic fatigue syndrome (Nater et al., 2008), major depressive disorder (Linkowski, 2003; Yehuda et al., 1996), and increased incidence and progression of disease (Sephton & Spiegel, 2003).

Dysregulation of circadian rhythms is often observed in patients suffering from psychological disorders, such as depression and schizophrenia. As opposed to just being a symptom of these disorders, it has been proposed that such dysregulation might contribute to their etiology (Karatsoreos, 2014; Menet & Rosbash, 2011). This is supported by accumulating evidence for SCN and circadian rhythm influence on mood and cognitive performance, including numerous demonstrations that impairment of the SCN in animal models has downstream consequences for various brain regions involved in cognition and emotion (for review, see Menet & Rosbash, 2011). Relevant to cortisol and the CAR, animal studies have shown that alterations to SCN integrity and function cause impairments to hippocampus-dependent memory (Ruby et al., 2008; Stephan & Kovacevic, 1978) and that such effects on memory can be induced by simply changing the pattern of the light/dark cycle (Devan et al., 2001). This potentially implicates the CAR, as a light-sensitive, time-of-day marker modulated by the SCN and HC and associated with HC-dependent memory and other cognitive functions.

An abnormal CAR is most often thought of as a side effect (or 'biomarker') of pathology, but it remains possible the CAR may contribute the relationship between chronic stress and impaired cognition. It is known that glucocorticoids entrain peripheral clocks in humans (Cuesta et al., 2015; Oster et al., 2016; Pezük, Mohawk, Wang, & Menaker, 2012), and this suggests that the circadian cortisol rhythm (of which the CAR is the most prominent aspect) may influence the temporal programmes of gene expression in diverse brain regions, complementing their neural regulation by circadian inputs (Herbert et al., 2006). It is also recognized that mismatch between the circadian cortisol and neural patterns may contribute to localized malfunction of these brain regions (Herbert et al., 2006).

In this context, and in light of the direct modulation of the adrenal cortex by the SCN (via the splanchnic nerve) to enhance adrenal sensitivity to ACTH in the immediate post-awakening period, it is reasonable to hypothesize that the CAR may function as a hormonal time-of-day marker and to influence the timing of peripheral clocks (Clow, Hucklebridge, Stalder, et al., 2010; Law et al., 2013). Such a hypothesis is currently speculative, but deserves further investigation.

5. The CAR and cognition

There are well established circadian rhythms in cognitive performance in humans, including declarative memory and executive function (Blatter & Cajochen, 2007; Dijk, Duffy & Czeisler, 1992; Schmidt, Collette, Cajochen, & Peigneux, 2007; Valdez, Reilly, & Waterhouse, 2008), and it is speculated that the CAR may play a role in the modulation of daytime cognition (see Clow, Hucklebridge, Stalder, et al., 2010; Fries et al., 2009). One hypothesis, proposed by Clow, Hucklebridge, Stalder, et al. (2010), is that the CAR might assist in the recovery of cognitive functions in the post-awakening period; a process known as 'sleep inertia' (SI; for review, see Hilditch and McHill, 2019). Electroencephalograph (EEG) and brain imaging studies indicate that although awakening from sleep comprises rapid reestablishment of consciousness, the attainment of alertness is relatively slow. SI is considered to be the time lag between these two states and has been shown to typically last anywhere between 1- and 30-min post-awakening (Hilditch and McHill, 2019). SI is therefore temporally associated with the CAR, as the substantial increase in cortisol has been shown to begin approximately 10 min post-awakening (shown in healthy young females; Smyth, Clow, Thorn, Hucklebridge, & Evans, 2013) and peak at around 30–45 min (Clow et al., 2004).

SI is known to be influenced by circadian phase and sleep stage upon awakening (Tassi and Muzet, 2000), and a proposed cause of SI is the delay in blood flow reaching the anterior cortical regions of the brain after awakening (Balkin et al., 2002). Scheer, Shea, Hilton, and Shea, (2008) used body temperature measurements to establish the circadian phase at the time of waking within a forced desynchrony protocol, and found that the worst SI impairment of cognition occurred when participants were woken during their biological 'night' (approximately between 2300 and 0300 h of the circadian cycle). Notably, it has also been demonstrated that the CAR is blunted in cases of night-time awakening (Dettenborn, Rosenloecher, & Kirschbaum, 2007), again indicating the potential for an association between these two processes. Finally, there is some evidence to suggest that exposure to light during this immediate post-waking period is an effective countermeasure for the symptoms of SI (Hilditch & McHill, 2019). This might again implicate the CAR in assisting with the recovery from SI, given that cortisol and the CAR are stimulated by light exposure (Scheer & Buijs, 1999; Thorn et al., 2004).

Numerous studies have indicated relationships between the CAR and indices of cognition beyond the immediate post-awakening period, including declarative memory, prospective memory, working memory, and executive functions. The results of these studies have been inconsistent, likely due to variation in nature and timing of both CAR and cognitive measures, different sample populations, and irregularity in monitoring sampling accuracy (see Stalder et al., 2016). This inconsistency of methodology and results, combined with the lack of reviews or meta-analyses in this area, limits understanding about CAR and cognition associations. It is the aim of the present review to address this limitation and shed some light on the associations between CAR and cognition. A summary of studies exploring the relationship between the CAR and cognition is presented in Table 1.

6. The CAR and memory functions

It is well established that cortisol secretory activity is associated with integrity of the HC (as described in Section 1). For example, circulating cortisol influences memory storage and retrieval (Lupien et al., 2009), and elevated circulating cortisol during sleep negatively impacts upon on simultaneous HC-dependent memory consolidation (Born & Wagner, 2009). The HC has been implicated in the regulation of hormonal responses to physical and psychological challenge, primarily in an inhibitory capacity, contributing to the negative feedback cycle (Fries et al., 2009; Herman, Ostrander, Mueller, & Figueiredo, 2005). The CAR too has been shown to be associated with HC integrity, volume, and associated function (Almela et al., 2012; Bruehl, Wolf, & Convit, 2009; Buchanan, Kern, Allen, Tranel, & Kirschbaum, 2004; Pruessner et al., 2007; Wolf, Fujiwara, Luwinski, Kirschbaum, & Markowitsch, 2005). For example, neither is the CAR observed in individuals with unilateral and bilateral HC lesions, nor those with severe global amnesia (Buchanan et al., 2004; Wolf et al., 2005). Further, it has been demonstrated that a larger CAR is associated with increased HC volume in healthy young men (Pruessner et al., 2007), and impaired declarative memory performance in older adults (Almela et al., 2012).

Several studies have explored associations between the CAR and HC-dependent memory. For example, CAR magnitude has been shown to be positively associated with episodic memory in healthy adults (Ennis et al., 2016). In a sample of elderly participants without dementia, Geerlings et al. (2015) also demonstrated that higher levels of cortisol at 45-min post-awakening are associated with better processing speed, spatial working memory and executive function (assessed by digits backward test and a variant of the Stroop task). While Geerlings et al. (2015) did not assess the CAR directly, the results implicate a possible role of the CAR due to the timing of the cortisol assessment (i.e., around the same time as the typical CAR peak), and would be in agreement with the accumulating evidence for a relationship between cortisol secretion in the post-awakening period and HC function. This study also indicated an association with slightly greater white matter volume, although there was no association with general memory performance or with volume of gray matter.

The association between the CAR and HC function may be bi-directional, such that the HC may regulate cortisol secretion during the CAR period, but so too may the increase in cortisol secretion during the CAR period influence hippocampal function. This is because not only does an abnormal CAR occur in the presence of lesions to these brain regions, but also that changes in cortisol secretion and the CAR influence HC-dependent memory functions. For example, Rimmele et al. (2010) demonstrated that pharmacologic suppression of the CAR using metyrapone impaired subsequent free recall of prior day learnt text and imagery.

The importance of HC function upon awakening is apparent, as it underlies not only episodic memory, but also spatial and temporal orientation (Burgess, Maguire, & O'Keefe, 2002; O'Keefe and Nadel, 1978). It has therefore been hypothesized that the CAR may play a role in activation of prospective memory function upon-awakening, encouraging the orientation of self in terms of space and time, and dealing with the cognitive demands of the post-awakening period (Bäumler, Voigt, et al., 2014; Bäumler, Kliegel, et al., 2014; Fries et al., 2009). This hypothesis was tested by Bäumler, Voigt, et al. (2014) who demonstrated that, in a sample of young children, performance on a game-like prospective memory task was positively associated with CAR magnitude. This was followed by a study within a (smaller) sample of the same young children which showed that a naturalistic prospective memory intervention (requiring the child to remind their parent of a planned gift upon awakening) resulted in a larger CAR compared to that on control days. As such, these findings not only provide support for the previously discussed CAR and HC-associated memory function relationship, but also for the hypothesis that the CAR is related to anticipation of challenge in the coming day (Clow, Hucklebridge, Stalder, et al., 2010; Fries et al., 2009).

In addition to declarative and prospective memory, the CAR has also been investigated in relation to working memory. A larger trait CAR (measured across 2 days) has been shown to be associated with increased working memory performance in older adult males in one study, though the same study failed to find a relationship in females of similar age (Almela et al., 2012). Moriarty et al. (2014) also reported a strong relationship between overall levels of cortisol secretion within the post-awakening period and spatial working memory in healthy older adults. The latter study indicated a U-shaped relationship, such that extreme high or low levels of cortisol were associated with decreased memory performance, which the authors interpreted as consistent with the previously reported positive relationship. However, this study did not employ electronic controls for participant sampling accuracy and the relationship was only observed with total cortisol secretion, as opposed to a (dynamic) CAR measure. Contradicting both of these reports, a recent study by Butler et al. (2017) found there to be no association between the CAR and working memory. In sum, evidence for a relationship between CAR and working memory can be considered tentative and inconsistent. The inconsistency in this area might suggest a need for further research, with appropriate application of electronic measures to monitor sapling accuracy, and careful control for other known confounds in CAR research (Clow, Hucklebridge, & Thorn, 2010; Stalder et al., 2016).

7. The CAR and executive function

One of the most prominent structures responsible for executive function is the frontal cortex (FC) which, like the HC, both expresses a high density of cortisol receptors and plays a role in regulating the HPA axis (Lupien et al., 2009). This relationship too is bi-directional, such that endogenous glucocorticoids modulate the cognitive functions of the pre-frontal cortex (Mizoguchi et al., 2004). However, relationships between cortisol secretion and executive function are complicated, not least because executive function is a broad and non-unitary concept encompassing many high-level cognitive processes, which depend upon several separate pre-frontal cortex structures. But also because many executive functions are responsible for control and organization of other cognitive functions (Diamond, 2013; Gilbert & Burgess, 2008; Miyake et al., 2000). While consensus has yet to be reached on an exact definition of executive function (Gilbert & Burgess, 2008), perhaps the best evidence is for there being three separate types: 'set shifting' (or 'cognitive flexibility') which is the ability to switch between competing task demands, 'updating' or working memory functions, and 'inhibition' which includes both cognitive and behavioral inhibitory processes (Diamond, 2013; Miyake et al., 2000).

Outside of the CAR literature, it has been demonstrated that acute increases in exogenous glucocorticoids are positively associated with set shifting and response inhibition measures (Dierolf et al., 2016; Shields et al., 2015; Vaz, Pradella-Hallinan, Bueno, & Pompeia, 2011). A recent study also has demonstrated a positive association between acute increases in endogenous cortisol and immediate, short-term enhancement of cognitive flexibility (Dierolf et al., 2016). However, studies in this field have yielded inconsistent results depending upon the type of executive function assessed and the relative timing of the measures, likely due to the differential effects of rapid (non-genomic), and slower (genomic) effects of cortisol secretion (Shields et al., 2015). For example, both Wingenfeld, Wolf, Krieg, and Lautenbacher (2011) and Vaz et al. (2011) have reported that such associations are no longer apparent if set shifting is measured more than 1 h after an acute increase in exogenous cortisol. There are good theoretical grounds for expecting rapid, non-genomic effects of cortisol to enhance response inhibition as this would support greater cognitive control over actions when dealing with a stressor (Shields et al., 2015), which would not necessarily be adaptive if experienced long after (at the genomic level).

Therefore, the criticality of both type and timing of executive function measures is apparent. This is a major complication for CAR-cognition studies also, and such differences in timing of the relevant measures might explain why there have been similarly inconsistent results in this field (see Table 1).

CAR magnitude has been shown to be positively associated with both working memory and attention-switching performance in older adults (Almela et al., 2012; Evans et al., 2012), as well as a measure of response inhibition in the afternoon of the same day in healthy young adults (Shi et al., 2018). Oosterholt et al. (2016) also found that in a sample of patients with clinical burnout, CAR magnitude was associated with better updating performance and fewer cognitive failures. But negative associations with task updating, speed of memory, error monitoring, and serial sequence learning have been reported in other samples (Hodyl et al., 2016; Maldonado et al., 2008; Oosterholt et al., 2016; Zhang et al., 2015), as well as one study reporting no association with task switching in healthy young adults (Butler et al., 2017). Similar inconsistency has occurred in studies of CAR associations with overall cognitive performance (e.g., Aas et al., 2011; Evans et al., 2011; Labad et al., 2016).

Associations between CAR and cognitive flexibility (set shifting) have been somewhat more consistent, with a positive association between CAR magnitude and cognitive flexibility (Trail Making task performance) found in older adults (Evans et al., 2012), and a positive association with attention-switching performance seen both in a sample of healthy young adults (Law et al., forthcoming) and a single case study of a healthy young male (Law et al., 2015). The positive association between cognitive flexibility and CAR magnitude was not observed in a recent study reporting a negative association between the CAR and emotion and gender task switching (Dierolf et al., 2016). However, as the CAR was defined by Dierolf et al. as mean morning cortisol AUCg across three consecutive days prior to the experiment (therefore not measuring the dynamic CAR per se), this finding might be better understood as a relationship between task switching and total post-awakening cortisol secretion.

A limitation for interpreting the findings of many CAR-EF studies is that they have not accounted for the substantial day-to-day variation in CAR magnitude (for review, see Law et al., 2013). The majority have been correlational between-subjects, or 'trait,' studies, often measuring the CAR on just one or 2 days. This is a limitation since it is thought day-to-day (state) differences account for around two thirds of the variance in CAR magnitude (Almeida, Piazza, & Stawski, 2009; Hellhammer et al., 2007; Stalder,

Hucklebridge, Evans, & Clow, 2009, Stalder et al., 2010b), and it has been recommended by Hellhammer et al. (2007) that to achieve a reliable trait measure of the CAR requires sampling over six consecutive days. In addition, such state variation is associated with a range of environmental and behavioral factors, such as differences in lighting and awakening times, that can potentially confound trait studies (Law et al., 2013). While the impact of daily variation in the CAR on EF has received less attention, the within-subject positive association between CAR and attention switching at 45 min postawakening (Law et al., 2015), as well as the within-subjects finding of a same-day association between CAR magnitude and an index of brain plasticity (Clow et al., 2014), offer some indication of the potential for investigating state associations between CAR and EF, even in healthy young participants.

8. Recommendations for future research

As described above, the CAR is the most prominent, dynamic and variable part of the circadian pattern of cortisol secretion, is under the modulatory influence of the SCN masterclock, and is timed to mark the significant circadian event of morning awakening. As such, there is good reason to hypothesize that one role for the CAR may be to act as a time-of-day marker for synchronization of circadian rhythms. Such a role for the CAR could potentially explain associations between CAR magnitude and the function of brain regions with a high affinity for cortisol, including memory and executive functions of the hippocampus and pre-frontal cortex, respectively. Future research exploring associations between the CAR and other cognitive functions related to these brain regions could further elucidate this relationship, while studies of clock gene function (using clock gene knockout mice, for example) might have potential to provide insight into the underlying mechanisms.

To ensure clarity and comparability of research findings, it will be important for future studies in this area to include careful controls for the many potentially confounding factors described here. For example, clear consideration and recording of the timing of cognitive measures in relation to the CAR is of critical importance in order to differentiate between rapid, non-genomic effects (for example, within 15–30 min of the CAR) and slower acting, genomic effects observed later in the day (Joëls et al., 2011). Of perhaps principle significance for valid CAR measurement is the inclusion of measures to ensure sampling accuracy, as laid out in the consensus guidelines of Stalder et al. (2016). While further consideration should also be given to day-to-day (state) differences in the CAR (as described in Law et al., 2013) as study of such variation has considerable potential to provide insight into CAR function, for example, by within-subject measurement of the CAR over several days along with same-day cognitive measures. Moreover, such state variation must be taken into account in future between-subject (trait) studies of the CAR by sampling over several days to achieve a reliable trait measure (Hellhammer et al., 2007). Implementing controls for potentially confounding state factors such as differences in awakening times and ambient light exposure might also allow for greater precision in determining associations between CAR and cognition (Law et al., 2013). Indeed, recent evidence indicates that blue/green wavelength light exposure is a more significant potential confounding factor than had previously been appreciated, though this also may have potential for investigation as an intervention for CAR manipulation (Petrowski et al., 2019).

If the CAR does support the re-establishment of cognitive functions (Clow, Hucklebridge, Stalder, et al., 2010; Fries et al., 2009; Law et al., 2013) then in theory, the light-responsive nature of the CAR would allow for optimally enhanced cognitive function when awakening at dawn, in preparation for the day ahead. A recent, detailed assessment of the implications of differing ambient light levels between CAR observations has indicated that post-awakening blue or green wavelength light influences CAR magnitude (Petrowski et al., 2019). This is a factor which can be easily controlled for either by direct measurement of light levels or, in ambulatory studies, simply by controlling for season, time of awakening and type of lighting (e.g., LED vs. fluorescent bulb; Petrowski et al., 2019). This, too, may present a direction for future research, as the recovery of executive functions such as cognitive flexibility may be assessed within the context of differences in ambient light at awakening. Indeed, preliminary support for such a relationship is provided by studies using functional magnetic resonance imaging techniques which indicate that exposure to light during the morning period results in increased activation of the subcortical regions associated with alertness and working memory (Vandewalle et al., 2006, 2010). Variation of light exposure therefore presents a potential method for manipulating CAR magnitude in order to examine effects on cognition or the activation of associated brain regions. Moving beyond correlational studies, experiments involving light-based intervention or other means for CAR manipulation might have considerable potential for further advancing understanding of the relationship between the CAR and cognition.

9. Conclusion

In summary, there is a growing body of research exploring relationships between stress, the CAR, and cognition. Within this literature, much of the focus has been on HC function, though there has been increasing interest in associations with frontal functions in recent years. Given the methodological difficulties within the area there have been relatively consistent demonstrations of correlational relationships in older adults: a larger average CAR being associated with better average memory and executive function. Day-to-day variation in the CAR has also been associated with day-to-day variation in executive function in healthy young populations However, there has been no dedicated effort to explore the mechanisms that might underpin a causal relationship between the CAR and cognition. It is suggested here that the CAR, which is regulated by two distinct pathways from the central body clock (SCN), might be a powerful time-of-day marker, synchronizing peripheral clocks in the brain to optimize brain function appropriate for the demands of the upcoming day. Chronic stress and aging lead to attenuation of the CAR and this may provide a route by which these factors impact cognition. If shown to be the case this provides a valuable target for intervention in cognitive decline. It is exciting to speculate that behavioral strategies that optimize the CAR (e.g., light, stress management, physical activity, early awakening, not smoking) might prove beneficial in a range of conditions not typically thought to be associated with neuroendocrine function. As such, there are many important questions in this field that remained unanswered and which present promising directions for further research.

References

Aas, M., Dazzan, P., Moondelli, V., Toulopoulou, T., Reichenberg, A., Di Forti, M., et al. (2011). Abnormal cortisol awakening response predicts worse cognitive function in patients with first-episode psychosis. *Psychological Medicine*, *41*(3), 463–476.

Adam, E. K., Hawkley, L. C., Kudielka, B. M., & Cacioppo, J. T. (2006). Day-to-day dynamics of experience–cortisol associations in a population-based sample of older adults. *Proceedings of the National Academy of Sciences of the United States of America*, *103*, 17058–17063.

Adam, E. K., Quinn, M. E., Tavernier, R., McQuillan, M. T., Dahlke, K. A., & Gilbert, K. E. (2017). Diurnal cortisol slopes and mental and physical health outcomes: A systematic review and meta-analysis. *Psychoneuroendocrinology*, *83*, 25–41.

Almela, M., van der Meij, L., Hidalgo, V., Villada, C., & Salvador, A. (2012). The cortisol awakening response and memory performance in older men and women. *Psychoneuroendocrinology*, *37*(12), 1929–1940.

Almeida, D. M., Piazza, J. R., & Stawski, R. S. (2009). Interindividual differences and intra-individual variability in the cortisol awakening response: An examination of age and gender. *Psychology and Aging*, *24*(4), 819–827.

Antle, M. C., & Silver, R. (2005). Orchestrating time: Arrangements of the brain circadian clock. *Trends in Neurosciences*, *28*(3), 145–151.

Balkin, T. J., Braun, A. R., Wesensten, N. J., Jeffries, K., Varga, M., Baldwin, P., et al. (2002). The process of awakening: A PET study of regional brain activity patterns mediating the re-establishment of alertness and consciousness. *Brain*, *125*(10), 2308–2319.

Bäumler, D., Kliegel, M., Kirschbaum, C., Miller, R., Alexander, N., & Stalder, T. (2014). Effect of a naturalistic prospective memory-related task on the cortisol awakening response in young children. *Biological Psychology*, *103*, 24–26.

Bäumler, D., Voigt, B., Miller, R., Stalder, T., Kirschbaum, C., & Kliegel, M. (2014). The relation of the cortisol awakening response and prospective memory functioning in young children. *Biological Psychology*, *99*, 41–46.

Benarroch, E. E. (2011). The melanopsin system: Phototransduction, projections, functions, and clinical implications. *Neurology*, *76*(16), 1422–1427.

Berson, D. M. (2007). Phototransduction in ganglion-cell photoreceptors. *Pflügers Archiv: European Journal of Physiology*, *454*(5), 849–855.

Berson, D. M., Dunn, F. A., & Takao, M. (2002). Phototransduction by retinal ganglion cells that set the circadian clock. *Science*, *295*(5557), 1070–1073.

Blatter, K., & Cajochen, C. (2007). Circadian rhythms in cognitive performance: Methodological constraints, protocols, theoretical underpinnings. *Physiology & Behavior*, *90*(2–3), 196–208.

Born, J., & Wagner, U. (2009). Sleep, hormones, and memory. *Obstetrics and Gynecology Clinics of North America*, *36*, 809–829.

Bornstein, S. R., Engeland, W. C., Ehrhart-Bornstein, M., & Herman, J. P. (2008). Dissociation of ACTH and glucocorticoids. *Trends in Endocrinology and Metabolism*, *19*(5), 175–180.

Bruehl, H., Wolf, O. T., & Convit, A. (2009). A blunted cortisol awakening response and hippocampal atrophy in type 2 diabetes mellitus. *Psychoneuroendocrinology*, *34*(6), 815–821.

Buchanan, T. W., Kern, S., Allen, J. S., Tranel, D., & Kirschbaum, C. (2004). Circadian regulation of cortisol after hippocampal damage in humans. *Biological Psychiatry*, *56*, 651–656.

Buchanan, T. W., & Lovallo, W. R. (2001). Enhanced memory for emotional material following stress-level cortisol treatment in humans. *Psychoneuroendocrinology*, *26*(3), 307–317.

Buijs, R. M., Eden, C. V., Goncharuk, V. D., & Kalsbeek, A. (2003). Circadian and seasonal rhythms—The biological clock tunes the organs of the body: Timing by hormones and the autonomic nervous system. *The Journal of Endocrinology*, *177*(1), 17–26.

Buijs, R. M., Wortel, J., Van Heerikhuize, J. J., & Kalsbeek, A. (1997). Novel environment induced inhibition of corticosterone secretion: Physiological evidence for a suprachiasmatic nucleus mediated neuronal hypothalamo-adrenal cortex pathway. *Brain Research*, *758*(1–2), 229–236.

Buijs, R. M., Wortel, J., Van Heerikhuize, J. J., Feenstra, M. G., Ter Horst, G. J., Romijn, H. J., et al. (1999). Anatomical and functional demonstration of a multisynaptic suprachiasmatic nucleus adrenal (cortex) pathway. *European Journal of Neuroscience*, *11*(5), 1535–1544.

Bunney, W. E., & Bunney, B. G. (2000). Molecular clock genes in man and lower animals: Possible implications for circadian abnormalities in depression. *Neuropsychopharmacology*, *22*(4), 335–345.

Burgess, N., Maguire, E. A., & O'Keefe, J. (2002). The human hippocampus and spatial and episodic memory. *Neuron*, *35*, 625–641.

Butler, K., Klaus, K., Edwards, L., & Pennington, K. (2017). Elevated cortisol awakening response associated with early life stress and impaired executive function in healthy adult males. *Hormones and Behavior, 95*, 13–21.

Chida, Y., & Steptoe, A. (2009). Cortisol awakening response and psychosocial factors: A systematic review and meta-analysis. *Biological Psychology, 80*(3), 265–278.

Clow, A., Hucklebridge, F., & Thorn, L. (2010). The cortisol awakening response in context. *International Review of Neurobiology, 93*, 153–175.

Clow, A., Hucklebridge, F., Stalder, T., Evans, P., & Thorn, L. (2010). The cortisol awakening response: More than a measure of HPA axis function. *Neuroscience and Biobehavioral Reviews, 35*, 97–103.

Clow, A., Law, R., Evans, P., Vallence, A. M., Hodyl, N. A., Goldsworthy, M. R., et al. (2014). Day differences in the cortisol awakening response predict day differences in synaptic plasticity in the brain. *Stress, 17*(3), 219–223.

Clow, A., Thorn, L., Evans, P., & Hucklebridge, F. (2004). The awakening cortisol response: Methodological issues and significance. *Stress, 7*, 29–37.

Cuesta, M., Cermakian, N., & Boivin, D. B. (2015). Glucocorticoids entrain molecular clock components in human peripheral cells. *The FASEB Journal, 29*(4), 1360–1370.

Cullen, A. E., Zunszain, P. A., Dickson, H., Roberts, R. E., Fisher, H. L., Pariante, C. M., et al. (2014). Cortisol awakening response and diurnal cortisol among children at elevated risk for schizophrenia: Relationship to psychosocial stress and cognition. *Psychoneuroendocrinology, 46*, 1–13.

De Kloet, E. R., Oitzl, M. S., & Joëls, M. (1999). Stress and cognition: Are corticosteroids good or bad guys? *Trends in Neurosciences, 22*(10), 422–426.

De Kloet, E. R., Van Acker, S. A., Sibug, R. M., Oitzl, M. S., Meijer, O. C., Rahmouni, K., et al. (2000). Brain mineralocorticoid receptors and centrally regulated functions. *Kidney International, 57*(4), 1329–1336.

Dettenborn, L., Rosenloecher, F., & Kirschbaum, C. (2007). No effects of repeated forced wakings during three consecutive nights on morning cortisol awakening responses (CAR): A preliminary study. *Psychoneuroendocrinology, 32*(8), 915–921.

Devan, B. D., Goad, E. H., Petri, H. L., Antoniadis, E. A., Hong, N. S., Ko, C. H., et al. (2001). Circadian phase-shifted rats show normal acquisition but impaired long-term retention of place information in the water task. *Neurobiology of Learning and Memory, 75*, 51–62.

Diamond, A. (2013). Executive functions. *Annual Review of Psychology, 64*, 135–168.

Dierolf, A. M., Arlt, L. E., Roelofs, K., Kölsch, M., Hülsemann, M. J., Schächinger, H., et al. (2016). Effects of basal and acute cortisol on cognitive flexibility in an emotional task switching paradigm in men. *Hormones and Behavior, 81*, 12–19.

Dijk, D. J., Duffy, J. F., & Czeisler, C. A. (1992). Circadian and sleep/wake dependent aspects of subjective alertness and cognitive performance. *Journal of Sleep Research, 1*(2), 112–117.

Doane, L. D., & Adam, E. K. (2010). Loneliness and cortisol: Momentary, day-to-day, and trait associations. *Psychoneuroendocrinology, 35*(3), 430–441.

Domes, G., Rothfischer, J., Reichwald, U., & Hautzinger, M. (2005). Inverted-U function between salivary cortisol and retrieval of verbal memory after hydrocortisone treatment. *Behavioral Neuroscience, 119*(2), 512.

Duan, H., Yuan, Y., Zhang, L., Qin, S., Zhang, K., Buchanan, T. W., et al. (2013). Chronic stress exposure decreases the cortisol awakening response in healthy young men. *Stress, 16*(6), 630–637.

Edwards, S., Clow, A., Evans, P., & Hucklebridge, F. (2001). Exploration of the awakening cortisol response in relation to diurnal cortisol secretory activity. *Life Sciences, 68*, 2093–2103.

Ennis, G. E., Moffat, S. D., & Hertzog, C. (2016). The cortisol awakening response and cognition across the adult lifespan. *Brain and Cognition*, *105*, 66–77.

Evans, P., Fredhoi, C., Loveday, C., Hucklebridge, F., Aitchison, E., Forte, D., et al. (2011). The diurnal cortisol cycle and cognitive performance in the healthy old. *International Journal of Psychophysiology*, *79*(3), 371–377.

Evans, P., Hucklebridge, F., Loveday, C., & Clow, A. (2012). The cortisol awakening response is related to executive function in older age. *International Journal of Psychophysiology*, *84*(2), 201–204.

Fehm, H. L., Klein, E., Holl, R., & Voigt, K. H. (1984). Evidence for extrapituitary mechanisms mediating the morning peak of plasma cortisol in man. *The Journal of Clinical Endocrinology & Metabolism*, *58*(3), 410–414.

Franz, C. E., O'Brien, R. C., Hauger, R. L., Mendoza, S. P., Panizzon, M. S., Prom-Wormley, E., et al. (2011). Cross-sectional and 35-year longitudinal assessment of salivary cortisol and cognitive functioning: the Vietnam Era twin study of aging. *Psychoneuroendocrinology*, *36*(7), 1040–1052.

Fries, E., Dettenborn, L., & Kirschbaum, C. (2009). The cortisol awakening response (CAR): Facts and future directions. *International Journal of Psychophysiology*, *72*(1), 67–73.

Geerlings, M. I., Sigurdsson, S., Eiriksdottir, G., Garcia, M. E., Harris, T. B., Gudnason, V., et al. (2015). Salivary cortisol, brain volumes, and cognition in community-dwelling elderly without dementia. *Neurology*, *85*(11), 976–983.

Gilbert, S. J., & Burgess, P. W. (2008). Executive function. *Current Biology*, *18*(3), R110–R114.

Gomez-Sanchez, E. P. (2010). The mammalian mineralocorticoid receptor: Tying down a promiscuous receptor. *Experimental Physiology*, *95*(1), 13–18.

Gomez-Sanchez, E. P. (2014). Brain mineralocorticoid receptors in cognition and cardiovascular homeostasis. *Steroids*, *91*, 20–31.

Heaney, J. L., Phillips, A. C., & Carroll, D. (2010). Ageing, depression, anxiety, social support and the diurnal rhythm and awakening response of salivary cortisol. *International Journal of Psychophysiology*, *78*(3), 201–208.

Hellhammer, J., Fries, E., Schweisthal, O., Schlotz, W., Stone, A., & Hagemann, D. (2007). Several daily measurements are necessary to reliably assess the cortisol rise after awakening: State-and trait components. *Psychoneuroendocrinology*, *32*, 80–86.

Herbert, J., Goodyer, I. M., Grossman, A. B., Hastings, M. H., De Kloet, E. R., Lightman, S. L., et al. (2006). Do corticosteroids damage the brain? *Journal of Neuroendocrinology*, *18*(6), 393–411.

Herman, J. P., Ostrander, M. M., Mueller, N. K., & Figueiredo, H. (2005). Limbic system mechanisms of stress regulation: Hypothalamo-pituitary-adrenocortical axis. *Progress in Neuro-Psychopharmacology and Biological Psychiatry*, *29*(8), 1201–1213.

Hidalgo, V., Almela, M., Pulopulos, M. M., & Salvador, A. (2016). Memory performance is related to the cortisol awakening response in older people, but not to the diurnal cortisol slope. *Psychoneuroendocrinology*, *71*, 136–146.

Hilditch, C. J., & McHill, A. W. (2019). Sleep inertia: Current insights. *Nature and Science of Sleep*, *11*, 155–165.

Hinkelmann, K., Muhtz, C., Dettenborn, L., Agorastos, A., Moritz, S., Wingenfeld, K., et al. (2013). Association between cortisol awakening response and memory function in major depression. *Psychological Medicine*, *43*(11), 2255–2263.

Hodyl, N. A., Schneider, L., Vallence, A. M., Clow, A., Ridding, M. C., & Pitcher, J. B. (2016). The cortisol awakening response is associated with performance of a serial sequence reaction time task. *International Journal of Psychophysiology*, *100*, 12–18.

Joëls, M., Fernandez, G., & Roozendaal, B. (2011). Stress and emotional memory: A matter of timing. *Trends in Cognitive Sciences*, *15*(6), 280–288.

Joëls, M., Karst, H., DeRijk, R., & de Kloet, E. R. (2008). The coming out of the brain mineralocorticoid receptor. *Trends in Neurosciences, 31*(1), 1–7.

Juster, R. P., Sindi, S., Marin, M. F., Perna, A., Hashemi, A., Pruessner, J. C., et al. (2011). A clinical allostatic load index is associated with burnout symptoms and hypocortisolemic profiles in healthy workers. *Psychoneuroendocrinology, 36*(6), 797–805.

Karatsoreos, I. N. (2014). Links between circadian rhythms and psychiatric disease. *Frontiers in Behavioral Neuroscience, 8*, 162.

Kirschbaum, C., Wolf, O. T., May, M., Wippich, W., & Hellhammer, D. H. (1996). Stress- and treatment-induced elevations of cortisol levels associated with impaired declarative memory in healthy adults. *Life Sciences, 58*(17), 1475–1483.

Knoops, A. J., van der Graaf, Y., Willem, P. T. M., & Geerlings, M. I. (2010). Age-related changes in hypothalamic–pituitary–adrenal axis activity in patients with manifest arterial disease. *Endocrine, 37*(1), 231–238.

Labad, J., Gutiérrez-Zotes, A., Creus, M., Montalvo, I., Cabezas, Á., Solé, M., et al. (2016). Hypothalamic-pituitary-adrenal axis measures and cognitive abilities in early psychosis: Are there sex differences? *Psychoneuroendocrinology, 72*, 54–62.

Law, R., Evans, P., Thorn, L., Hucklebridge, F., & Clow, A. (2015). The cortisol awakening response predicts same morning executive function: Results from a 50-day case study. *Stress, 18*(6), 616–621.

Law, R., Evans, P., Thorn, L., Hucklebridge, F., Loveday, C., & Clow, A. The cortisol awakening response predicts a same-day index of executive function in healthy young adults. *International Journal of Psychophysiology*, forthcoming.

Law, R., Hucklebridge, F., Thorn, L., Evans, P., & Clow, A. (2013). State variation in the cortisol awakening response. *Stress, 16*(5), 483–492.

Linkowski, P. (2003). Neuroendocrine profiles in mood disorders. *The International Journal of Neuropsychopharmacology, 6*(2), 191–197.

Linkowski, P., Mendlewicz, J., Leclercq, R., Brasseur, M., Hubain, P., Golstein, J., et al. (1985). The 24-hour profile of adrenocorticotropin and cortisol in major depressive illness. *The Journal of Clinical Endocrinology and Metabolism, 61*(3), 429–438.

Lovell, B., Moss, M., & Wetherell, M. A. (2011). Perceived stress, common health complaints and diurnal patterns of cortisol secretion in young, otherwise healthy individuals. *Hormones and Behavior, 60*(3), 301–305.

Lupien, S., Lecours, A. R., Lussier, I., Schwartz, G., Nair, N. P., & Meaney, M. J. (1994). Basal cortisol levels and cognitive deficits in human aging. *Journal of Neuroscience, 14*(5), 2893–2903.

Lupien, S. J., McEwen, B. S., Gunnar, M. R., & Heim, C. (2009). Effects of stress throughout the lifespan on the brain, behaviour and cognition. *Nature Review Neuroscience, 10*, 434–445.

Maina, G., Palmas, A., Bovenzi, M., & Filon, F. L. (2009). Salivary cortisol and psychosocial hazards at work. *American Journal of Industrial Medicine, 52*(3), 251–260.

Maldonado, E. F., Fernandez, F. J., Trianes, M. V., Wesnes, K., Petrini, O., Zangara, A., et al. (2008). Cognitive performance and morning levels of salivary cortisol and α-amylase in children reporting high vs. low daily stress perception. *The Spanish Journal of Psychology, 11*(1), 3–15.

McEwen, B. S. (1999). Stress and hippocampal plasticity. *Annual Review of Neuroscience, 22*, 105–122.

McEwen, B. S. (2008). Central effects of stress hormones in health and disease: Understanding the protective and damaging effects of stress and stress mediators. *European Journal of Pharmacology, 583*(2), 174–185.

McEwen, B. S. (2012). Brain on stress: How the social environment gets under the skin. *Proceedings of the National Academy of Sciences of the United States of America, 109*(2), 17180–17185.

McEwen, B. S., & Gianaros, P. J. (2011). Stress-and allostasis-induced brain plasticity. *Annual Review of Medicine*, *62*, 431–445.

McEwen, B. S., & Sapolsky, R. M. (1995). Stress and cognitive function. *Current Opinion in Neurobiology*, *5*(2), 205–216.

Menet, J. S., & Rosbash, M. (2011). When brain clocks lose track of time: Cause or consequence of neuropsychiatric disorders. *Current Opinion in Neurobiology*, *21*(6), 849–857.

Miller, G. E., Chen, E., & Zhou, E. S. (2007). If it goes up, must it come down? Chronic stress and the hypothalamic-pituitary-adrenocortical axis in humans. *Psychological Bulletin*, *133*(1), 25.

Miyake, A., Friedman, N. P., Emerson, M. J., Witzki, A. H., Howerter, A., & Wager, T. D. (2000). The unity and diversity of executive functions and their contributions to complex "frontal lobe" tasks: A latent variable analysis. *Cognitive Psychology*, *41*(1), 49–100.

Mizoguchi, K., Ishige, A., Takeda, S., Aburada, M., & Tabira, T. (2004). Endogenous glucocorticoids are essential for maintaining prefrontal cortical cognitive function. *The Journal of Neuroscience*, *24*(24), 5492–5499.

Moriarty, A. S., Bradley, A. J., Anderson, K. N., Watson, S., Gallagher, P., & McAllister-Williams, R. H. (2014). Cortisol awakening response and spatial working memory in man: A U-shaped relationship. *Human Psychopharmacology: Clinical and Experimental*, *29*(3), 295–298.

Nater, U. M., Youngblood, L. S., Jones, J. F., Unger, E. R., Miller, A. H., Reeves, W. C., et al. (2008). Alterations in diurnal salivary cortisol rhythm in a population-based sample of cases with chronic fatigue syndrome. *Psychosomatic Medicine*, *70*(3), 298–305.

O'Connor, D. B., Hendrickx, H., Dadd, T., Elliman, T. D., Willis, T. A., Talbot, D., et al. (2009). Cortisol awakening rise in middle-aged women in relation to psychological stress. *Psychoneuroendocrinology*, *34*(10), 1486–1494.

O'Keefe, J., & Nadel, L. (1978). *The Hippocampus as a Cognitive Map*. .

Oskis, A., Loveday, C., Hucklebridge, F., Thorn, L., & Clow, A. (2009). Diurnal patterns of salivary cortisol across the adolescent period in healthy females. *Psychoneuroendocrinology*, *34*(3), 307–316.

Oosterholt, B. G., Maes, J. H., Van der Linden, D., Verbraak, M. J., & Kompier, M. A. (2016). Getting better, but not well: A 1.5 year follow-up of cognitive performance and cortisol levels in clinical and non-Clinical burnout. *Biological Psychology*, *117*, 89–99.

Oster, H., Challet, E., Ott, V., Arvat, E., de Kloet, E. R., Dijk, D. J., et al. (2016). The functional and clinical significance of the 24-hour rhythm of circulating glucocorticoids. *Endocrine Reviews*, *38*(1), 3–45.

Petrowski, K., Schmalbach, B., Niedling, M., & Stalder, T. (2019). The effects of post-awakening light exposure on the cortisol awakening response in healthy male individuals. *Psychoneuroendocrinology*, *108*, 28–34.

Pezük, P., Mohawk, J. A., Wang, L. A., & Menaker, M. (2012). Glucocorticoids as entraining signals for peripheral circadian oscillators. *Endocrinology*, *153*(10), 4775–4783.

Pittenger, C., & Duman, R. S. (2008). Stress, depression, and neuroplasticity: A convergence of mechanisms. *Neuropsychopharmacology*, *33*(1), 88–109.

Pruessner, J. C., Hellhammer, D. H., & Kirschbaum, C. (1999). Burnout, perceived stress, and cortisol responses to awakening. *Psychosomatic Medicine*, *61*(2), 197–204.

Pruessner, J. C., Wolf, O. T., Hellhammer, D. H., Buske-Kirschbaum, A., Von Auer, K., Jobst, S., et al. (1997). Free cortisol levels after awakening: A reliable biological marker for the assessment of adrenocortical activity. *Life Sciences*, *61*(26), 2539–2549.

Pruessner, M., Pruessner, J. C., Hellhammer, D. H., Pike, G. B., & Lupien, S. J. (2007). The associations among hippocampal volume, cortisol reactivity, and memory performance in healthy young men. *Psychiatry Research: Neuroimaging*, *155*(1), 1–10.

Rimmele, U., Meier, F., Lange, T., & Born, J. (2010). Suppressing the morning rise in cortisol impairs free recall. *Learning & Memory*, *17*(4), 186–190.

Ruby, N. F., Hwang, C. E., Wessells, C., Fernandez, F., Zhang, P., Sapolsky, R., et al. (2008). Hippocampal-dependent learning requires a functional circadian system. *Proceedings of the National Academy of Sciences of the United States of America*, *105*, 15593–15598.

Saban, K. L., Mathews, H. L., Bryant, F. B., O'Brien, T. E., & Janusek, L. W. (2012). Depressive symptoms and diurnal salivary cortisol patterns among female caregivers of stroke survivors. *Biological Research for Nursing*, *14*(4), 396–404.

Salvat-Pujol, N., Labad, J., Urretavizcaya, M., de Arriba-Arnau, A., Segalàs, C., Real, E., et al. (2017). Hypothalamic-pituitary-adrenal axis activity and cognition in major depression: The role of remission status. *Psychoneuroendocrinology*, *76*, 38–48.

Scheer, F. A., & Buijs, R. M. (1999). Light affects morning salivary cortisol in humans. *The Journal of Clinical Endocrinology and Metabolism*, *84*(9), 3395–3398.

Scheer, F. A., Shea, T. J., Hilton, M. F., & Shea, S. A. (2008). An endogenous circadian rhythm in sleep inertia results in greatest cognitive impairment upon awakening during the biological night. *Journal of Biological Rhythms*, *23*(4), 353–361.

Schilling, T. M., Kölsch, M., Larra, M. F., Zech, C. M., Blumenthal, T. D., Frings, C., et al. (2013). For whom the bell (curve) tolls: Cortisol rapidly affects memory retrieval by an inverted U-shaped dose–response relationship. *Psychoneuroendocrinology*, *38*(9), 1565–1572.

Schmidt, C., Collette, F., Cajochen, C., & Peigneux, P. (2007). A time to think: Circadian rhythms in human cognition. *Cognitive Neuropsychology*, *24*(7), 755–789.

Schmidt-Reinwald, A., Pruessner, J. C., Hellhammer, D. H., Federenko, I., Rohleder, N., Schürmeyer, T. H., et al. (1999). The cortisol response to awakening in relation to different challenge tests and a 12-hour cortisol rhythm. *Life Sciences*, *64*(18), 1653–1660.

Sephton, S., & Spiegel, D. (2003). Circadian disruption in cancer: A neuroendocrine-immune pathway from stress to disease? *Brain, Behavior, and Immunity*, *17*(5), 321–328.

Shields, G. S., Bonner, J. C., & Moons, W. G. (2015). Does cortisol influence core executive functions? A meta-analysis of acute cortisol administration effects on working memory, inhibition, and set-shifting. *Psychoneuroendocrinology*, *58*, 91–103.

Shi, X., Sun, X., Yao, Z., Yuan, Y., Wu, J., & Clow, A. (2018). The cortisol awakening response predicts response inhibition in the afternoon of the same day. *Psychoneuroendocrinology*, *89*, 23–29.

Smyth, N., Clow, A., Thorn, L., Hucklebridge, F., & Evans, P. (2013). Delays of 5-15 minutes between awakening and the start of saliva sampling matter in assessment of the cortisol awakening response. *Psychoneuroendocrinology*, *38*(9), 1476–1483.

Smyth, N., Thorn, L., Hucklebridge, F., Clow, A., & Evans, P. (2016). Assessment of the cortisol awakening response: Real-time analysis and curvilinear effects of sample timing inaccuracy. *Psychoneuroendocrinology*, *74*, 380–386.

Stalder, T., Bäumler, D., Miller, R., Alexander, N., Kliegel, M., & Kirschbaum, C. (2013). The cortisol awakening response in infants: Ontogeny and associations with development-related variables. *Psychoneuroendocrinology*, *38*(4), 552–559.

Stalder, T., Evans, P., Hucklebridge, F., & Clow, A. (2010a). Associations between psychosocial state variables and the cortisol awakening response in a single case study. *Psychoneuroendocrinology*, *35*(2), 209–214.

Stalder, T., Evans, P., Hucklebridge, F., & Clow, A. (2010b). State associations with the cortisol awakening response in healthy females. *Psychoneuroendocrinology*, *35*(8), 1245–1252.

Stalder, T., Hucklebridge, F., Evans, P., & Clow, A. (2009). Use of a single case study design to examine state variation in the cortisol awakening response: Relationship with time of awakening. *Psychoneuroendocrinology*, *34*(4), 607–614.

Stalder, T., Kirschbaum, C., Kudielka, B. M., Adam, E. K., Pruessner, J. C., Wüst, S., et al. (2016). Assessment of the cortisol awakening response: Expert consensus guidelines. *Psychoneuroendocrinology*, *63*, 414–432.

Stawski, R. S., Almeida, D. M., Lachman, M. E., Tun, P. A., Rosnick, C. B., & Seeman, T. (2011). Associations between cognitive function and naturally occurring daily cortisol during middle adulthood: Timing is everything. *The Journals of Gerontology. Series B, Psychological Sciences and Social Sciences*, *66*, i71–i81.

Stephan, F. K., & Kovacevic, N. S. (1978). Multiple retention deficit in passive avoidance in rats is eliminated by suprachiasmatic lesions. *Behavioral Biology*, *22*(4), 456–462.

Tassi, P., & Muzet, A. (2000). Sleep inertia. *Sleep Medicine Reviews*, *4*(4), 341–353.

Thorn, L., Hucklebridge, F., Esgate, A., Evans, P., & Clow, A. (2004). The effect of dawn simulation on the cortisol response to awakening in healthy participants. *Psychoneuroendocrinology*, *29*, 925–930.

Valdez, P., Reilly, T., & Waterhouse, J. (2008). Rhythms of mental performance. *Mind, Brain, and Education*, *2*(1), 7–16.

Vandewalle, G., Balteau, E., Phillips, C., Degueldre, C., Moreau, V., Sterpenich, V., et al. (2006). Daytime light exposure dynamically enhances brain responses. *Current Biology*, *16*(16), 1616–1621.

Vandewalle, G., Schwartz, S., Grandjean, D., Wuillaume, C., Balteau, E., Degueldre, C., et al. (2010). Spectral quality of light modulates emotional brain responses in humans. *Proceedings of the National Academy of Sciences*, *107*(45), 19549–19554.

Vaz, L. J., Pradella-Hallinan, M., Bueno, O. F. A., & Pompeia, S. (2011). Acute glucocorticoid effects on the multicomponent model of working memory. *Human Psychopharmacology: Clinical and Experimental*, *26*(7), 477–487.

Vedhara, K., Hyde, J., Gilchrist, I. D., Tytherleigh, M., & Plummer, S. (2000). Acute stress, memory, attention and cortisol. *Psychoneuroendocrinology*, *25*(6), 535–549.

Vyas, A., Mitra, R., Rao, B. S. S., & Chattarji, S. (2002). Chronic stress induces contrasting patterns of dendritic remodelling in hippocampal and amygdaloid neurons. *Journal of Neuroscience*, *22*, 6810–6818.

Weitzman, E. D., Fukushima, D., Nogeire, C., Roffwarg, H., Gallagher, T. F., & Hellman, L. (1971). Twenty-four hour pattern of the episodic secretion of cortisol in normal subjects. *The Journal of Clinical Endocrinology and Metabolism*, *33*(1), 14–22.

Wilhelm, I., Born, J., Kudielka, B. M., Schlotz, W., & Wüst, S. (2007). Is the cortisol awakening rise a response to awakening? *Psychoneuroendocrinology*, *32*, 358–366.

Wingenfeld, K., Wolf, S., Krieg, J. C., & Lautenbacher, S. (2011). Working memory performance and cognitive flexibility after dexamethasone or hydrocortisone administration in healthy volunteers. *Psychopharmacology*, *217*(3), 323–329.

Wolf, O. T., Fujiwara, E., Luwinski, G., Kirschbaum, C., & Markowitsch, H. J. (2005). No morning cortisol response in patients with severe global amnesia. *Psychoneuroendocrinology*, *30*(1), 101–105.

Wright, C. E., & Steptoe, A. (2005). Subjective socioeconomic position, gender and cortisol responses to waking in an elderly population. *Psychoneuroendocrinology*, *30*(6), 582–590.

Wüst, S., Federenko, I., Hellhammer, D. H., & Kirschbaum, C. (2000). Genetic factors, perceived chronic stress, and the free cortisol response to awakening. *Psychoneuroendocrinology*, *25*(7), 707–720.

Wüst, S., Wolf, J., Hellhammer, D. H., Federenko, I., Schommer, N., & Kirschbaum, C. (2000). The cortisol awakening response-normal values and confounds. *Noise & Health*, *2*(7), 79–88.

Yehuda, R., Harvey, P. D., Buchsbaum, M., Tischler, L., & Schmeidler, J. (2007). Enhanced effects of cortisol administration on episodic and working memory in aging veterans with PTSD. *Neuropsychopharmacology*, *32*(12), 2581–2591.

Yehuda, R., Teicher, M. H., Trestman, R. L., Levengood, R. A., & Siever, L. J. (1996). Cortisol regulation in posttraumatic stress disorder and major depression: A chronobiological analysis. *Biological Psychiatry*, *40*(2), 79–88.

Zhang, L., Duan, H., Qin, S., Yuan, Y., Buchanan, T. W., Zhang, K., et al. (2015). High cortisol awakening response is associated with impaired error monitoring and decreased post-error adjustment. *Stress*, *18*(5), 561–568.

CHAPTER NINE

Effectiveness of stress-relieving strategies in regulating patterns of cortisol secretion and promoting brain health

Nina Smyth[a,*], Elena Rossi[a], Carly Wood[b]
[a]Psychology, School of Social Sciences, University of Westminster, London, England
[b]School of Life Sciences, University of Westminster, London, England
*Corresponding author: e-mail address: smythn@westminster.ac.uk

Contents

Abstract

Stress leads to ill-health and disease, and with today's fast-pace western society, engaging in strategies to relieve stress is crucial for good health across the life-course. Activities such as focusing on positive characteristics, art/music therapies, mindfulness, yoga and engaging with nature and/or physical activity have been shown to reduce stress and enhance well-being. It is thought that patterns of cortisol secretion, which are regulated by the brain, are a key mediator of stress-disease and well-being-health links. Measurement of cortisol in saliva is a non-invasive and ecologically valid tool for detecting early changes in brain health, as well as evaluating the effectiveness of strategies in relieving stress and improving brain health as well as monitoring stress-related

International Review of Neurobiology, Volume 150
ISSN 0074-7742
https://doi.org/10.1016/bs.irn.2020.01.003

brain changes. This chapter will review the evidence that engaging in stress-relieving strategies promotes regulation and/or restoration of patterns of cortisol secretion. If such strategies are found to be effective in healthy populations, they could potentially inform ways of promoting brain health and the prevention or delay of clinical disorders involving disorders in the brain (e.g., Parkinson's disease) and symptoms experienced with such disorders. To inform this field of research, recommendations are provided for the use of salivary cortisol as a marker of early monitoring of brain health and effectiveness of stress-alleviating interventions.

1. Stress-disease and well-being-health links

Ill-being (i.e., stress, negative affect, depression) and well-being (i.e., positive emotions, psychological good health, meaning in life) are related dimensions; but importantly they are distinct. The absence of ill-being does not necessarily mean the presence of well-being; an individual might not have meaning or purpose in their lives, which is important for good health (e.g., Diener & Emmons, 1984). Links between stress-disease and well-being-health are independent of each other and pathways mediating the associations have been shown to differ (e.g., Chida & Steptoe, 2008; Ryff et al., 2006).

Stress results in ill-health and early death; these links are clear and well documented (for reviews, see Cuijpers & Smit, 2002; Hemingway & Marmot, 1999; Kubzansky & Kawachi, 2000). A striking example of this comes from the Whitehall studies; British civil servants were studied over 10-year periods, and individuals in lower grade occupations reported the least control in their work demands and decision-making, the most effort-reward imbalance and injustice at work. These individuals were at a higher risk of ill-health, for example, higher risk of coronary heart disease and mortality (e.g., Kuper & Marmot, 2003; Marmot, Rose, Shipley, & Hamilton, 1978). Moreover, human flourishing has been shown to be just as, if not, more important for healthy functioning. A wealth of prospective, empirical and population based research reveals benefits of well-being on health, for example, lower mortality, reduced morbidity and pain, and better self-reported health lower risk of future coronary heart disease and reduced mortality in healthy and clinical samples (Chida & Steptoe, 2008; Kubzansky & Thurston, 2007; Pressman & Cohen, 2005). The study of Catholic nuns' diaries and smile intensity in photographs provide illustrative examples of the well-being-health link; positive emotional content written in the nuns' diaries and the smile intensity of baseball players predicted longevity (Abel & Kruger, 2010; Danner, Snowdon, & Friesen, 2001). In light of this evidence, both ill-being and well-being need consideration in order to promote healthy functioning.

2. Cortisol: Its role in health

Cortisol is secreted during situations perceived as stressful via the hypothalamic-pituitary-adrenal (HPA) axis; higher brain regions activate the hypothalamus to secrete corticotrophin releasing hormone (CRH), triggering the release of adrenocorticotrophic hormone (ACTH) by the anterior pituitary gland, which in turn stimulates the release of cortisol from the adrenal cortex. Secretion is regulated by negative feedback; cortisol receptors in multiple brain regions detect high cortisol levels, enabling individuals to adapt and recover from stress. More importantly, cortisol regulates 24-h function of down-stream physiological systems, controlled by the hypothalamic suprachiasmatic nucleus. A healthy pattern is characterized by a rapid increase of cortisol immediately after morning awakening, peaking about 30 min (known as the cortisol awakening response: CAR), followed by steadily declining levels thereafter (known as the diurnal decline), reaching the lowest levels in the evening/early sleep.

This dynamic cortisol pattern becomes dysregulated with chronic HPA axis activation, which is turn results in dysregulation of multiple down-stream physiological processes, accelerating ill-health (see Smyth, Hucklebridge, Thorn, Evans, & Clow, 2013). For example, a recent systematic review of stress responding shows that heightened and blunted responding to acute stressors in the laboratory are implicated in ill-health (Turner et al., 2020). Furthermore, a meta-analysis showed that flatter diurnal cortisol slopes over the day were predictive of a range of physical and mental ill-health outcomes, such as, cancer, depression, cardiovascular disease (Adam et al., 2017) and blunted CAR has been linked with aberrant brain health (Law & Clow, 2020; Shi et al., 2018). Thus evidence shows that heightened and blunted (i.e., lower) stress responding as well as high or low cortisol secretion and flattened cortisol patterns over the day are implicated in psychological and physical ill-health and cognitive deficits. However, it important to consider sex differences in the stress response system (see Smyth et al., 2013).

3. Cortisol "a window on the brain": A marker of brain health

Assessment of cortisol provides a "window on the brain"; cortisol secretion during stressful situations has wide effects on brain structure and function and the underlying 24-h cortisol circadian rhythm has a role in regulating other bodily systems, including brain function (Clow & Smyth, 2020). Measurement of cortisol in saliva is non-invasive, and it is relatively easy to collect making repeated sampling from individuals in their natural environments, going about their daily lives possible. Despite its relative ease in collection, accurate measurement and consideration of variables influencing cortisol is crucial to understanding stress-disease and well-being-health links (see Smyth et al., 2013). Effects of acute stress can be measured by the examination of cortisol responses (i.e., cortisol reactivity and recovery) to stressors induced under laboratory conditions. While the effects of chronic stress can be better examined by measurement of underlying basal cortisol circadian patterns.

Measuring patterns of cortisol secretion can provide a tool for enhancing and restoring brain health in healthy and clinical populations by allowing us to:

(i) *Detect early changes in brain health*: sustained activation of the HPA axis leads to dysregulated cortisol circadian rhythms, and deviations from "normal" cortisol secretory patterns can be detected before manifestation of aberrant brain function and disease.

(ii) *Promote brain health throughout the life-course*: identification of aberrant patterns of cortisol secretion (either too small or large responding to acute stressors or flattened diurnal cortisol patterns) can encourage strategies to restore patterns of cortisol secretion and in turn promote brain health.

(iii) *Evaluate effectiveness of stress-relieving activities in promoting health*: evaluate the effectiveness of stress-relieving activities in regulating and/or restoring patterns of cortisol secretion, and associated changes in cognition, stress/well-being and health. Cortisol has potent neurotoxic effects and restoring cortisol levels/patterns in those with aberrant levels/patterns (e.g., in brain disorder patients such as in Parkinson's disease), are likely to slow down and/or delay symptoms associated with the disorder (e.g., improvements in cognitive function which is influenced by cortisol patterns).

3.1 Cortisol circadian rhythms: Chronic stress exposure

3.1.1 The cortisol awakening response (CAR)

The CAR is typically measured by collecting saliva on awakening and every 15 min up to 45–60 min post-awakening. Collecting samples at the desired time, particularly the awakening sample and consideration of state variables that influence it, is key to its accurate measurement (Law, Hucklebridge, Thorn, Evans, & Clow, 2013; Smyth, Thorn, Hucklebridge, Clow, & Evans, 2016; Stalder, Hucklebridge, Evans, & Clow, 2009). The CAR is typically calculated by the mean rise (e.g., calculated as the mean increase: MnInc; area under the curve with respect to increase: AUCi; or the difference between the awakening and peak value, see Smyth et al., 2013). This measure of rise is different to overall levels in the post-awakening period (i.e., calculate mean of the four samples). However, more sophisticated growth curve modeling enables examination of patterns of cortisol secretion without the need to rely on composite measures, and this considered the optimal way to analyze cortisol data (see Smyth et al., 2016).

Cortisol secretion is pulsatile with bursts approximately every hour. It is considered the first "ultradian episode of the day"; it is the point at which the brain switches from sleeping to waking, kick-starting the ultradian rhythm. Thus it is suggested to provide additional insight into brain functioning than the profile over the rest of the day (Clow & Smyth, 2020; Evans, Smyth, Thorn, Hucklebridge, & Clow, 2019). The CAR is thought to prepare our body to deal with upcoming demands of the day (Adam, Hawkley, Kudielka, & Cacioppo, 2006) and has huge inter- and intra-variability; for example, a case study of a healthy male showed a range of 3.6 and 39.0 nmol/L over 50 days (Stalder et al., 2009). This points to the flexibility of the CAR to deal with daily demands (i.e., mounting a bigger rise in cortisol in the post-awakening period on more demanding days). However, most studies measure the CAR on 1 or 2 days limiting our understanding of its role and function. To obtain a "trait" measure of the CAR, it is recommended to measure it up to 6-days (Hellhammer et al., 2007), but this can be demanding and costly. Measuring CAR on repeated days would enable examination of flexibility of the response, and how this links with health. Recently, we showed that the shape of the CAR might be more indicative of health status than simply examining the size of the response. A significant burst of cortisol in the immediate awakening period (i.e., within 20 min) followed by a steep decline in levels (termed CAR Salience) has been shown to be a better measure of CAR and health status (Evans et al., 2019). It is important to understand that flatter/blunted CAR is different from not exhibiting a CAR (i.e., a negative CAR: no rise in cortisol). It is unclear whether a negative CAR is an indicator of health status and requires further investigation but consideration of this issue is needed in analyses (see Clow & Smyth, 2020).

3.2 The cortisol diurnal profile

Cortisol secretion declines steadily over the day. It is typically measured by the diurnal cortisol decline; the decline from the morning (but outside of the CAR period) through to the evening as well as mean cortisol levels over the day (i.e., mean levels over relevant sample points: mean diurnal cortisol). Since awakening times vary, sample collection should be anchored to awakening time (i.e., 3, 6, 9 and 12h post-awakening) and not clock time. It is essential to report both mean diurnal cortisol and diurnal cortisol decline since a flat cortisol pattern can be associated with either high or low cortisol secretion, and both are predictive of ill-health (Adam et al., 2017). More recently, consistency in the decline and mean levels of cortisol over days (i.e., less variation) has been shown to be important for good health (Segerstrom, Sephton, & Westgate, 2017), but measurement of the profile over several days is required to examine this.

Most studies examining cortisol circadian patterns do so only by measurement of it on 1 or 2 days and restricting analyses to computing simple composite measures, thus missing valuable opportunity to examine variability of cortisol as well as drilling down on the dynamic aspects of cortisol secretion over multiple days. Moreover, the majority of studies lack longitudinal analysis, for example, as little as 10% of studies (see Adam et al., 2017), thus limiting our understanding of its mediating role in stress-health and well-being-health.

3.3 Stress/cortisol responding

Stressors can be induced under laboratory conditions; the Trier Social Stress Test (TSST: Kirschbaum, Pirke, & Hellhammer, 1993) includes an anticipation element, social-evaluation and uncontrollability, and these elements have been shown to most reliably activate the HPA axis (Dickerson & Kemeny, 2004). The TSST requires participants to prepare for a speech which they deliver to an unsympathetic authoritarian audience followed by a mental arithmetic task. Under laboratory conditions, this task most resembles psychosocial stress experienced in today's society. Of course, there are other types of protocols (e.g., Stroop color-word interference task or mirror drawing) but they do not reliably activate the HPA axis (see Wetherell, Craw, & Smith, 2018). Cortisol levels peak about 20–30 min following onset of the stressor, returning to baseline in the following hour; to capture this wave of activity, saliva samples are collected prior to the stressor, and every 10–15 min up to 60-min from onset of the stressor (see Smyth et al., 2013). Both heightened and blunted responding are associated with a range of adverse behavioral and health outcomes in clinical and healthy populations. However, there are no normative values for a "healthy" response, and some individuals do not respond to the stressor. It is unclear if non-responding is a trait characteristic and/or a marker of health status or simply due to timing of the stressor onset. This issue requires further investigation but should be considered in analyses and reported (see Clow & Smyth, 2020).

Despite a wealth of research showing exaggerated and blunted responding is linked with ill-health, very few studies have examined the causal effects of stress/well-being in the etiology of good health and disease, as very few studies have examined negative effects of stress reactivity longitudinally (Turner et al., 2020). Moving away for longitudinal studies another way to examine causal effects is via intervention studies; however, repeated measures of exposing participants to stressors can be problematic due to habituation effects (see Smyth et al., 2013). Our preliminary, unpublished data show that if exposure to a stressor (e.g., TSST) is induced following a significant period (i.e., 6 months to 1 year) exposure to the stressor the second time still activates the HPA axis. Thus, assigning participants to stress-relieving interventions and measuring stress reactivity at baseline and at follow-up could be a way of untangling causal effects.

4. Stress-relieving strategies

Today's fast past society has promoted engagement in stress-relieving strategies as a way of reducing stress and enhancing well-being. Engaging in international activities, such as focusing on positive characteristics, art/music activities, mindfulness, yoga and engaging with nature and/or physical activity, have been shown to be effective in reducing stress and enhancing well-being. The Social Prescribing initiative, in the United Kingdom, enables individuals suffering from stress-related problems and brain disorders to be referred to such activities via community and charity initiatives. For example, Neurologists work as part of a multidisciplinary team of health care professionals and some work with the Social Prescribing link worker model (see Fixsen & Marie, forthcoming). Evidence-based research on stress-relieving strategies in healthy and clinical population, especially those with brain disorders will promote wider scale roll out and take up of Social Prescribing in clinical and non-clinical settings.

Examination of patterns of cortisol secretion, measured in saliva, is the ideal was to evaluate the effectiveness of stress-relieving interventions as chronic stress results in dysregulation of patterns of cortisol secretion, and this is an early marker of brain function dysregulation. Thus, it provides a "window on the brain" in a non-invasive and ecological valid way. The next section of this chapter provides an overview of some of the evidence demonstrating the effectiveness of intentional activities in healthy regulation and/or restoration of the patterns of cortisol secretion.

4.1 Positive psychological interventions

The positive psychology movement alerted our attention to ways of enhancing well-being through engaging in intentional activities aimed at focusing on positive emotions, engagement, relationships, meaning, and accomplishments. A randomized-controlled trial of positive psychological interventions (such as expressing gratitude, identifying and using personal strengths or writing about the good things in your life, or you at your best) in adults (N=577, aged 35–54 years) revealed significant increases in happiness immediately and at longer follow-up points; with individuals in the gratitude condition showing the largest boosts in well-being, sustained for up to 1 month (Seligman, Steen, Park, & Peterson, 2005). A growing body of evidence demonstrates the effectiveness of positive psychological interventions in enhancing well-being (Boehm, Lyubomirsky, & Sheldon, 2011), but future research will need to determine if these techniques have beneficial effects on brain function, potentially via regulation of patterns of cortisol secretion.

5. Arts and music therapies

The last few decades have witnessed the growth and consideration of artistic therapies such as music or visual art making/observation for management of stress; with overall beneficial effects of art therapies on stress-related outcomes being reported (Martin et al., 2018). Differing effects on the type of music has been observed for cortisol secretion; relaxing music reduces cortisol while stimulating music has been shown to increase it (Fancourt, Ockelford, & Belai, 2014). Studies have revealed anxiolytic effects of music both during and post-surgical procedures in a variety of patients. For instance, Noriya Uedo et al. (2004) observed smaller increases in salivary cortisol concentrations in patients undergoing screening colonoscopy exposed to relaxing music compared to those who underwent the surgery without music. Kar, Ganguly, Roy, and Goswami (2015) revealed similar results in patients undergoing surgery for cardiopulmonary bypass, with significantly lower secretion of cortisol secretion in those receiving the procedure while listening to music. Mottahedian, Sahraei, Movahhedi, Hajizadeh, and Lak (2012) also found smaller increases of cortisol levels in post-operative patients exposed to music during surgery in comparison to control groups.

Other artistic activities have been explored in association with reduced stress. For instance, Clow and Fredhoi (2006) investigated the effect of a short lunchtime visit to an art gallery on outcomes of psychological and physiological stress in London City workers. Results showed a marked decrease in self-reported levels of stress as well as cortisol levels after the art gallery visit. Importantly, the study revealed that cortisol levels reduced significantly only for those participants whose baseline levels were already much higher than "normative" values prior to the art gallery visit. This suggests a particular role of such activities on the stress system where the latter tends to present over-active, and has important implications for enforcing these artistic activities/breaks to lower and maintain normal levels of stress, both psychologically and physiologically. Other studies on art making (e.g., drawing, painting) also reported decreasing self-reported levels of stress following being exposed to a stressor (color-word inference test and mental arithmetic test) compared with non-artistic tasks; however, no physiological measures were obtained (Abbott, Shanahan, & Neufeld, 2013). Although results suggest that artistic activities can lead to emotional and physiological changes with beneficial effects for health and well-being, more research is needed with attention to physiological markers such as cortisol, as well as increasing the length of interventions. This research is required to increase the current knowledge on how best to introduce art-making activities in people's lives, either through therapy or simply attending sessions; and tackle with a preventative approach to stress-related disorders and promotion of brain health.

5.1 Mindfulness-based interventions

Mindfulness, an act of consciously focusing one's attention on the present moment, in a patient and non-judgmental manner, has received a great amount of attention in recent decades. Mindfulness-based Stress Reduction (MBSR) was one of the first clinical applications, originally implemented to alleviate suffering (Kabat-Zinn, 1982); and its application was extended to reduce stress and alleviate symptoms in those with ill-health (Grossman, Niemann, Schmidt, & Walach, 2004). A wealth of research shows the effectiveness of mindfulness in reducing stress and enhancing well-being; however, there is less evidence on the effectiveness in restoring cortisol patterns and brain health.

A systematic review of MBSR interventions (29 studies; $N=2668$ healthy adults) observed large effects for reductions in stress (Khoury, Sharma, Rush, & Fournier, 2015). One study measured cortisol status in adults who reported increased stress; cortisol reactivity to a psychosocial stressor (the TSST) was compared between those assigned to an 8-week MBSR intervention or waiting list control condition. No changes in cortisol concentrations between conditions or pre- and post-intervention were found (Nyklíček, Mommersteeg, Van Beugen, Ramakers, & Van Boxtel, 2013), although there was a limited number of saliva samples collected and thus the peak of cortisol might not have been captured. Lower cortisol 30 min following awakening was observed in teachers who completed an 8-week MBSR intervention compared to those assigned to the control group (Flook, Goldberg, Pinger, Bonus, & Davidson, 2013). Goldberg et al. (2014) found decreased hair cortisol levels following a mindfulness-based intervention in those engaged in smoking cessation treatment. Measuring cortisol in hair can be problematic since it provides a trait measure of cortisol secretion (average over 1–3 months), therefore only providing average levels and not the dynamic pattern over the day. This is problematic because both high and low cortisol levels have been associated with ill-health and it does not say anything about the pattern. High levels of cortisol could indicate high levels in the post-awakening period and declining levels over the day, which is a healthy profile, or it could mean the opposite: a flatter profile with low levels in the post-awakening period and higher levels over the reminder of the day. It is therefore important to consider these "limitations" when interpreting the data.

Mindfulness-based interventions conducted with clinical populations have also been shown to be effective in reducing self-reported stress from pre- to post-intervention and in comparison, to control groups (Mackenzie, Carlson, & Speca, 2005; Ott, Norris, & Bauer-Wu, 2006; Smith, Richardson, Hoffman, & Pilkington, 2005), and some studies show sustained effects for up to 6-months (e.g., Carlson, 2018). Much of the research has been conducted on cancer patients, an ideal group to study as they experience high levels stress and aberrant cortisol profiles (e.g., Ryan et al., 2017). Mindfulness-based interventions were most effective in reducing cortisol levels for breast cancer patients who displayed the highest cortisol levels pre-intervention (Carlson, Speca, Patel, & Goodey, 2004). Sustained effects of reducing cortisol levels were observed for up to 1-year; however, effects were less pronounced for those who had been living with cancer for a longer period (Carlson, Speca, Faris, & Patel, 2007). There is support for lower cortisol concentrations following mindfulness-based interventions in other clinical groups; for example, lower plasma cortisol in participants with diabetes mellitus (Jung, Lee, & Park, 2015), but in participants with inactive ulcerative colitis there were no changes in urinary cortisol concentration (Jedel et al., 2014). Like hair cortisol, urinary cortisol provides a measure of trait cortisol secretion over a 24-h period, and not the dynamic pattern over the day. Mindfulness-based interventions have also been shown to restore cortisol circadian patterns; for example, a steeper cortisol profile over the day was observed in cancer patients randomly allocated to a mindfulness-based intervention (Carlson et al., 2013). Evidence on mindfulness-based practice as a stress-relieving activity is promising but more rigorous cortisol methodology and focus on examining changes in diurnal cortisol patterns and associated changes in stress/well-being and cognition is needed to better understand its role in brain health. Moreover, focus on a wider range of clinical brain conditions would enable examination of how changes in patterns of cortisol secretion are associated with clinical symptoms (e.g., improvements in cognition).

5.2 Yoga practice

Yoga practice, a combination of physical postures, breathing techniques, relaxation and meditation, has increased in popularity in recent years with numerous health benefits reported (Cartwright, Mason, Porter, & Pilkington, 2020). For example, yoga practice has been shown to reduce stress; in a systematic review of 35 studies, 25 provided support for the efficiency of yoga practice in reducing self-reported stress (Li & Goldsmith, 2012). However, fewer studies examined the impact of yoga practice on patterns of cortisol secretion (only 8 out of 35 studies), and those that did provide conflicting evidence. For instance, the majority of the studies did not report any changes in cortisol following yoga practice (Beddoe, Yang, Kennedy, Weiss, & Lee, 2009; Bosch, Traustadóttir, Howard, &

Matt, 2009; Schell, Allolio, & Schonecke, 1994). In the studies that did observe lower cortisol levels following yoga practice this was for acute changes in cortisol secretion; with lower salivary cortisol following 90-min yoga practice in emotionally distressed females, college students and pregnant depressed (Field, Diego, Delgado, & Medina, 2013; Michalsen et al., 2005; West, Otte, Geher, Johnson, & Mohr, 2004). However, changes in cortisol levels were not always associated with changes in psychosocial measures (e.g., West et al., 2004). Few studies have examined the efficiency of yoga programs over several weeks or months; however, Rocha et al. (2012) reported that healthy Military males ($N=36$ aged 20–40 years) had lower levels of cortisol in the morning (collected at 7:00) following two 60-min weekly classes for 6-months and in comparison to the control group. Older adults ($N=118$, aged 62 years) randomly assigned to three 60-min weekly classes over 8-weeks also had lower cortisol reactivity to a stressor (completion of a cognitive task) following the intervention and in comparison to the control group, and lower cortisol levels predicted better cognitive performance for the yoga group while increases in cortisol was predictive of worse performance in the controls (Gothe, Keswani, & McAuley, 2016). In contrast, Granath, Ingvarsson, von Thiele, and Lundberg (2006) did not observe reductions in cortisol concentrations following a 4-month yoga intervention in adults reporting stress-related problems; however, salivary cortisol was only measured once 2 weeks before and after intervention.

Support for regulation of cortisol in clinical populations is mixed; no changes in urinary or salivary cortisol concentrations following a 12 week yoga program (two 70-min weakly sessions for first 6 weeks followed by once a week for the remaining 6-weeks) were found in hypertension patients (Cohen et al., 2011). However, salivary cortisol was measured between 7:00 and 20:00 and authors did not state if time of day was consistent for pre- and post-salivary measures; this is a limitation as cortisol levels change over the day thus timing of its measurement needs to be consistent form pre- to post-measurement. Reduced cortisol concentrations in the morning and evening (collected at 6:00, 9:00 and 21:00) were observed for breast cancer patients undergoing radiotherapy ($N=44$) assigned to yoga intervention versus the control condition (Raghavendra et al., 2009) and in breast cancer patients practicing yoga twice weekly for 8-weeks (Banasik, Williams, Haberman, Blank, & Bendel, 2011). Different yoga practices, such as Iyengar yoga (a type of yoga that gets progressively more difficult as strength and flexibility increase), compared, for example, to the more classic Hatha or Ashtanga yoga, could differentially contribute to modulation of the

HPA axis. Unfortunately, not all studies report the type of yoga practice being implemented, perhaps limiting a full interpretation of the findings. Additionally, studies should include an active control group (rather than the typical wait-list), which is important to exclude general effects of the activity on the outcomes measured.

Much of the evidence on the benefits of Yoga is based on average levels of cortisol, but examining cortisol circadian patterns, particularly the CAR, provides a better measure of brain-health. In individuals suffering from Fibromyalgia there were no differences in the CAR or cortisol diurnal decline, only average cortisol levels, following engagement in an 8-week yoga program (Curtis, Osadchuk, & Katz, 2011). However, sample size was small ($N=19$) and cortisol patterns were only measured on 2 days; suggesting that a larger number of participants, and longer measurements are key understanding effects of yoga. Likewise, in a non-clinical MBSR intervention, yoga practice did not result in changes in CAR, diurnal decline, mean levels or serum cortisol levels despite measure of salivary cortisol over 4 days pre- and post-intervention (Daubenmier et al., 2011). It is suggested that yoga practice might be most effective in restoring the CAR in those chronically stressed who are exhibited blunted or exaggerated CARs (Anderson & Wideman, 2017). However, it is difficult to make firm conclusions given the limited number of studies examining the CAR and yoga practice.

A meta-analysis of yoga programs (all randomized trails) revealed reductions in waking, afternoon and evening cortisol levels pre- to post-intervention as well as in comparison with the control groups; but sensitivity analysis suggested these reductions were moderate. Moreover, no differences in cortisol levels at 30 min or 60 min post-awakening or mid-morning, or cortisol diurnal decline were reported (Pascoe, Thompson, & Ski, 2017). Overall the evidence on yoga practice as a stress-relieving activity is mixed mostly because of poor methodology. Future research examining the effects of yoga practice prior to exposure to a stressor in the laboratory would inform its ability to manage responses and recovery from acute stress while measurement of the diurnal cortisol pattern prior to and following participation in yoga practice over several months would inform its ability to regulate and/or restore cortisol circadian patterns.

5.3 Physical exercise

Being physically active is associated with healthier CAR profiles. For example, engagement in exercise restored a "blunted CAR" to a "healthy" and "normal" CAR in depressed individuals (Vreeburg et al., 2009) and in police officers who experience high stress (Fekedulegn et al., 2018). Puterman et al. (2011) examined the effects of being physically active on stress responding in 46 post-menopausal women who underwent the TSST. Results clearly outlined the role of physical activity; sedentary participants with high levels of rumination exhibited a rapid increase of cortisol during the test, and delayed recovery of cortisol levels after the test. On the contrary, a relationship between rumination levels and cortisol patterns was absent in physically active participants.

Research in "healthy groups" has also highlighted the role of physical exercise in reducing stress reactivity. Klaperski, von Dawans, Heinrichs, and Fuchs (2013) reported significantly increased cortisol reactivity to the TSST in sedentary individuals, compared to moderately and vigorously active individuals. Furthermore, a 12-week endurance exercise program resulted in significantly reduced cortisol reactivity to the TSST (Klaperski, von Dawans, Heinrichs, & Fuchs, 2014). These changes were significantly different to those experienced by the waiting control and relaxation groups; however, the relaxation group also experienced a reduction in cortisol reactivity following the relaxation intervention. Notably, the exercise group were the only group to experience changes in their physical fitness over the course of the intervention; suggesting a potential role of fitness in reducing stress reactivity. This is supported by the findings of Wood, Clow, Hucklebridge, Law, and Smyth (in review) who reported reduced overall cortisol responses to the TSST in "fit" compared to "unfit" individuals. However, much of the evidence on the effects of physical fitness is focused on athletic populations and does not use gold-standard methods of fitness assessment such as the VO_2max test; limiting the interpretation of findings to the wider population.

Moderate to higher-intensive physical activity is thought to prevent or delay age-related cognitive deficits via regulation of brain activity and physiological processes (Paterson & Murias, 2014), such as regulation of patterns of cortisol secretion. For example, being physically active was associated with better cognitive function; similarly, a steeper cortisol diurnal decline was associated with better physical performance and cognitive function in adults with and without mild cognitive impairment (Dijckmans et al., 2017). However, the authors could not determine the mediating role of cortisol in the relationship between physical activity and cognitive function. In individuals with mild cognitive impairment, a 3-month exercise program resulted in greater decline in cortisol from awakening to midday compared to the control group, but there was no difference in the CAR; additionally, improvements were seen in some aspects of cognitive and executive functions but not memory (Tortosa-Martínez et al., 2015). A strength of this study was that the salivary cortisol post intervention was collected 5 days after the last exercise session, to obtain accurate stable cortisol values and to avoid reporting acute effects of physical activity on cortisol levels. Smyth et al. (2019) examined the effects of a 6-month high intensity endurance exercise program on participants with Parkinson's disease with mild cognitive impairment. Overall levels of cortisol were lower across the day, but mostly attributed to the morning, and these levels were more in line with levels exhibited by healthy adults of a similar age. Although there were a lack of observed differences in the dynamic profile of cortisol (i.e., CAR or diurnal decline); this might be due to measuring cortisol only on 2 days or it might be that aberrant cortisol patterns in Parkinson's disease patients are attributed to hyper-secretion of cortisol and not changes in the dynamic of the profile. While this was a small study ($N=8$), it was an intensive study and the gold standard salivary cortisol methodology was fully adhered to (i.e., electronic monitoring of saliva sampling, full measurement of cortisol circadian patterns with samples anchored to waking time); strengthening the validity of the findings.

Aerobic exercise is a form of low intensity/impact; there is mixed evidence for the relationship between aerobic exercise and cortisol patterns. For example, a 12-week aerobic or stretching program resulted in improvements in memory for depressed patients, but results for the CAR were mixed with the aerobic group exhibiting a smaller CAR and the stretching group exhibiting a bigger CAR following the program (Foley et al., 2008). However, this was a small sample ($N=23$) and there was large inter-individual variability in the CAR, which was measured only on 1 day with no control of sampling accuracy, meaning the data may be unreliable. More recently, a 6-month aerobic exercise program resulted in restoration of a blunted CAR to a bigger increase post intervention, in healthy older adults, but changes in cortisol did not predict changes in cognition (Drogos et al., 2019). In addition, dance or movement training, as a light exercise, have been shown to be effective in reducing cortisol levels over the day, but not the dynamic of cortisol in healthy older adults (Vrinceanu et al., 2019); while a steeper cortisol slope over the day was found for breast cancer patients following dance movement training (Ho, Fong, & Yip, 2018). Similarly, in Dementia patients ($N=204$) who were randomly allocated to 10-weeks of dance movement therapy or a matched intensity exercise condition or the control group, a steeper cortisol slope over the day was observed in the dance movement group in comparison with participants in the control group, and effects were sustained at 1-year follow-up. Improvements in mood and aspects of cognitive function (i.e., retrieval and delayed recall) were also observed for participants in the dance therapy group in comparison to the control group, but no significant improvements between the exercise conditions and the control group (Ho, Fong, Chan, et al., 2018).

Of interest, single bouts of exercise are unlikely to change the CAR; two studies did not provide support for changes in CAR when exercise was carried out the day prior to the CAR measurement (Garde et al., 2009; Uçar, Özgöçer, & Yildiz, 2018). However, single bouts of exercise might be effective in reducing overall cortisol levels; low-level physical exercise reduced cortisol levels in young adults on the Autistic Spectrum, who are considered to experience severe stress and anxiety (Hillier, Murphy, & Ferrara, 2011).

Evidence on physical activity in promoting brain health is more advanced than other activities reviewed, mainly because more rigors research has been conducted. Many studies point to the effectiveness of physical activity in managing responses to stress as well as regulating and restoring cortisol circadian patterns. Moreover, this evidence comes from both healthy and clinical populations, and what is exciting is that evidence points to changes in patterns of cortisol secretion is related with other improvements in brain health, such as cognitive function. To further advance this area longitudinal analysis of physical activity in healthy populations and intervention studies involving a range of brain disorders is needed using optimal measurement of cortisol alongside measurement of improvements in health and/or clinical symptoms.

5.4 Nature and physical activity

Exposure and access to natural environments reduces stress and is associated with many health benefits (Barton, Bragg, Wood, & Pretty, 2016). Earlier studies revealed that salivary cortisol levels were lower for participants who were shown scenes of nature or exposed to natural environments (e.g., walking in a forest) compared to those exposed to urban scenes or environments (e.g., Beil & Hanes, 2013; Lee, Park, Tsunetsugu, Kagawa, & Miyazaki, 2009; Park et al., 2007). Exposure to a stressor in the laboratory, a stressful version of the Stroop Test, resulted in lower cortisol levels in allotment gardeners that were randomly assigned to outdoor gardening compared to those who read a book indoors (Van Den Berg & Custers, 2010). Although results are promising, most of these studies are conducted on small samples; nonetheless, support was provided for reduced cortisol levels from a larger study of 280 participants who visited forests or city environments on different days; salivary cortisol was lower when they visited the forest environments compared to when they visited the city areas (Park, Tsunetsugu, Kasetani, Kagawa, & Miyazaki, 2009).

Growing evidence supports healthier cortisol circadian rhythms in those who have more access to natural environments. For example, more green-space in the neighborhood was associated with a steeper cortisol decline over the day in unemployed adults ($N=25$, aged 35–55 years) living in deprived areas. Although they exhibited cortisol levels within the expected "healthy" range, they exhibited a flattened profile, with low levels of cortisol in the morning and little decline over the day, meaning higher levels later in the day. This highlights the importance of examining diurnal patterns over the day since average levels of cortisol were within the expected range but by examining day patterns, we can see that levels are low when they should be higher, and vice versa (Ward Thompson et al., 2012). To extend these findings, Roe et al. (2013) examined the importance of neighborhood green space in a larger sample ($N=106$) and observed differences in this relationship for males and females; females with little green space in their neighborhood, showed a low and flat cortisol slope (i.e., low levels of cortisol over the day with little decline from morning to the evening) while males with little green-space in their neighborhood showed a high and flat cortisol profile (i.e., high levels over the day with little decline from morning to evening). These findings highlight the need to examine males and females separately as we cannot assume that stress-relieving activities will have the same effects for males and females.

Access and exposure to natural environments encourages physical activity, and physical activity in natural environments (i.e., Green Exercise) provides additional benefits for health and well-being than nature exposure or physical activity alone (see Barton et al., 2016). A systematic review of 43 studies using a variety of designs (i.e., comparison between participants allocated to different conditions or comparing participants who completed all conditions on different days or weeks), across a range of countries and settings demonstrated the benefits of exposure to nature and green exercise in reducing stress and cortisol concentrations, with natural environments having the most benefits in reducing stress (Kondo, Jacoby, & South, 2018). But changes in cortisol were at times independent of changes in psychosocial variables and sometimes no changes in cortisol were observed whilst changes in psychosocial measures were. Fewer studies have examined green exercise and cortisol circadian rhythms but a work-based green-exercise (25 min bike session and 20 min strength session) program was shown to restore the CAR in 14 office-workers; CAR was heightened, outside of the expected range prior to the intervention and was lower post-intervention compared to those that exercised indoors (Calogiuri et al., 2015).

Recently, we conducted a study to examine the efficiency of an outdoor walk in regulating our response to stress. Healthy young females completed a 30 min walk prior to being exposed to the group version of the TSST; and exhibited significantly lower levels of cortisol in comparison to healthy females who did not go for a walk (Wood et al., in review). No differences were observed for cortisol reactivity; however, this pilot study did not account for cortisol non-responders. Furthermore, the walk consisted of walking through an urban area to access a green environment, thus we could not determine if a nature-based walk was more beneficial than a walk in a non-nature environment as the exact effects of the green exercise could not be determined. To untangle the effects of green-exercise on stress, we conducted a well-controlled and intensive study where 18 healthy adults completed a 30-min treadmill exercise at 50% of their $V0_2max$ while viewing a nature, built or blank screen. Participants in the urban exercise conditions exhibited a "blunted" cortisol response and reported higher stress. Although a small study ($N = 18$), exercise intensity was based on individual's fitness levels ensuring that exercise intensity was carefully controlled. Results need to be confirmed in a larger sample, and when exercise is performed in a natural environment rather than while viewing a natural environment (Wood, Flynn, Law, Naufahu, & Smyth, in review).

Natural environments might encourage physical activity, but with the increasing urban development access to these are decreasing and our connections with nature are changing. Thus, we need to consider individual differences in our connections with nature and the barriers to accessing it when considering the health benefits of nature and/or physical activity (Gladwell, Brown, Wood, Sandercock, & Barton, 2013). Furthermore, we showed that engaging in green exercise in childhood predicted exposure and connectedness to nature in adulthood, and those with reduced stress (measured by heart rate variability) were more connected to nature (Wood & Smyth, 2019). This finding unveils the importance of our connectedness to nature and promotes a prevention approach to intervene and educate early in childhood on the health benefits of nature and/or physical activity, which could greatly impact adult health and tackle adverse outcomes of stress.

The evidence base on engaging with nature and green exercise suggests its effective in managing our responses to stress as well as promoting healthier cortisol circadian rhythms but this evidence is limited to cross-sectional studies. Longitudinal studies examining the long-term effects of these activities is needed as well as larger scale intervention studies where exercise intensity and fitness level and our connection to nature are considered as well as the effect of social interaction, since much of these activities contain social interaction which if enjoyed is associated with healthier patterns of cortisol secretion. Studies to date have failed to examine the contributing effects of the components involved (i.e., engagement/exposure to nature, physical activity and social interaction).

5.5 The importance of enjoyment and engagement in stress-relieving activities

Motivation to engage with an activity and enjoyment of that activity might promote engagement in stress-relieving activities and these factors may impact the effectiveness of the activity in reducing stress, enhancing well-being and restoring and/or regulating patterns of cortisol secretion. For instance, enjoyment in physical activity predicts peoples engagement in physical activity (Lewis, Williams, Frayeh, & Marcus, 2016) and enjoyment predicts exercise level (Hagberg, Lindahl, Nyberg, & Hellénius, 2009). Unfortunately, the majority of studies examining the effectiveness of stress-relieving strategies, did not measure these factors nor did they consider them as influencing factors. These variables could change over the course of participation in activities, thus it is important to measure these factors throughout participation in these activities to understand how they influence effectiveness of relieving stress and promoting well-being. It is also important to note that there might be individual differences in these variables and thus one activity might not be suitable for all; engaging in different activities to explore how these activities work for the individual and settling on one that is most enjoyable is likely to have the most health-related benefits. For instance, preliminary unpublished results from our group have indicated that participants' enjoyment of a range of intentional activities designed to relieve stress was a key predictor of beneficial changes in patterns of cortisol secretion in healthy adults.

6. Conclusions

Evidence for the efficiency of intentional activities in alleviating stress and enhancing well-being, and in turn promoting health is promising. However, the evidence for patterns of cortisol secretion mediating the relationship between stress-disease and well-being-health is mixed; inconsistencies in results might be driven by differences in methodology, insufficient power due to small samples, study design, lack of control groups and/or randomization, and type of analysis. The majority of studies are limited to examining changes in cortisol concentrations, but both high and low cortisol levels are linked with dysregulated cortisol circadian rhythms and ill-health. Fewer studies have examined the efficiency of intentional activities in regulating and/or restoring our responses and recovery from acute stress or cortisol circadian rhythms, and those that do fail to employ rigorous measurement of cortisol that is essential for its accurate measurement. Thus, this field of research is still in its infancy and has a long way ahead before we can determine the efficiency of stress-relieving activities in the regulation of patterns of cortisol secretion and in turn healthy functioning. Understanding how intentional activities can promote the prevention and/or delay of brain disorders as well as the management of clinical symptoms requires longitudinal studies across healthy populations and patients with brain disorders and examining associated changes in health/well-being and brain health (e.g., cognitive function). We have made the following recommendations for future research:

- Individuals who drop out of the intervention study might be those most in need of engaging in stress-relieving activities; patterns of cortisol secretion at baseline, where possible, should be compared with those who did not drop out to understand who these interventions are most effective for.
- Measure cortisol in saliva samples since this enables measurement of stress responding or diurnal cortisol patterns over the day. While other measures (i.e., urine or hair) provide trait average levels and the dynamic patterns cannot be determined.
- The consensus CAR guidelines must be adhered to for accurate measurement of the CAR (Stalder et al., 2016).
- Saliva sampling is a novel activity (can produce larger CAR on the first sampling day) and the CAR is influenced by state variables; thus measurement of CAR on 1 or 2 days is not sufficient to obtain a baseline measure of CAR. Between 3 and 6 days would provide a better measure of "trait" CAR for measuring changes in CAR pre- and post-intervention.

- Measuring patterns of cortisol secretion on repeated days will also enable examination of appropriate variability of cortisol secretion (i.e., secretion matched to daily demands).
- Examine the cortisol rise and decline in the post-awakening period (i.e., CAR salience) and not just the size of the rise (i.e., mean rise from awakening to peak).
- Examine both mean diurnal cortisol and cortisol decline (outside of the CAR period) as flattened profiles can be characterized by high and low cortisol levels, both associated with ill-health.
- Examine if patterns of cortisol secretion differ for males and females.
- Model the dynamic curve of cortisol (see Smyth et al., 2016) instead of using composite measures of cortisol (e.g., MnInc, AUCi), which can be influenced by inaccurate sampling.
- It is unclear whether stress non-responding and negative CARs are an indicator of health status or due to methodological issues; we should therefore report percentages of non-responders/negative CARs, and account for these in analyses.
- Several studies do not report associations between cortisol measures and self-report measures of stress and other relevant psychosocial variables; we should measure both as either alone might fail to capture accurate reflection of individuals state.
- Measure motivation, engagement and enjoyment levels throughout the intervention, including prior to participation in the stress-relieving activity.
- Longitudinal evaluation of stress-relieving activities is required to determine the efficiency of activities in preventing stress-related changes in brain health.
- Examine if changes in patterns of cortisol secretion are related with improvements in stress and well-being, and cognitive functioning. In individuals with brain disorders also examine if changes are related with reductions and/or delay of symptoms.
- Combine cortisol measurement and health/well-being assessment alongside other ways of brain imaging.

References

Abbott, K. A., Shanahan, M. J., & Neufeld, R. W. J. (2013). Artistic tasks outperform non-artistic tasks for stress reduction. *Art Therapy*, *30*(2), 71–78. https://doi.org/10.1080/07421656.2013.787214.

Abel, E. L., & Kruger, M. L. (2010). Smile intensity in photographs predicts longevity. *Psychological Science*, *21*(4), 542–544. https://doi.org/10.1177/0956797610363775.

Adam, E. K., Hawkley, L. C., Kudielka, B. M., & Cacioppo, J. T. (2006). Day-to-day dynamics of experience–cortisol associations in a population-based sample of older adults. *Proceedings of the National Academy of Sciences of the United States of America*, *103*(45), 17058–17063. https://doi.org/10.1073/pnas.0605053103.

Adam, E. K., Quinn, M. E., Tavernier, R., McQuillan, M. T., Dahlke, K. A., & Gilbert, K. E. (2017). Diurnal cortisol slopes and mental and physical health outcomes: A systematic review and meta-analysis. *Psychoneuroendocrinology*, *83*, 25–41. https://doi.org/10.1016/j.psyneuen.2017.05.018.

Anderson, T., & Wideman, L. (2017). Exercise and the cortisol awakening response: A systematic review. *Sports Medicine—Open*, *3*(1), 37. https://doi.org/10.1186/s40798-017-0102-3.

Banasik, J., Williams, H., Haberman, M., Blank, S. E., & Bendel, R. (2011). Effect of Iyengar yoga practice on fatigue and diurnal salivary cortisol concentration in breast cancer survivors. *Journal of the American Academy of Nurse Practitioners*, *23*(3), 135–142. https://doi.org/10.1111/j.1745-7599.2010.00573.x.

Barton, J., Bragg, R., Wood, C., & Pretty, J. (2016). *Green exercise: Linking nature, health and well-being*. Routledge.

Beddoe, A. E., Yang, C.-P. P., Kennedy, H. P., Weiss, S. J., & Lee, K. A. (2009). The effects of mindfulness-based yoga during pregnancy on maternal psychological and physical distress. *Journal of Obstetric, Gynecologic & Neonatal Nursing*, *38*(3), 310–319. https://doi.org/10.1111/j.1552-6909.2009.01023.x.

Beil, K., & Hanes, D. (2013). The influence of urban natural and built environments on physiological and psychological measures of stress—A pilot study. *International Journal of Environmental Research and Public Health*, *10*(4), 1250–1267. https://doi.org/10.3390/ijerph10041250.

Boehm, J. K., Lyubomirsky, S., & Sheldon, K. M. (2011). A longitudinal experimental study comparing the effectiveness of happiness-enhancing strategies in Anglo Americans and Asian Americans. *Cognition and Emotion*, *25*(7), 1263–1272. https://doi.org/10.1080/02699931.2010.541227.

Bosch, P. R., Traustadóttir, T., Howard, P., & Matt, K. S. (2009). Functional and physiological effects of yoga in women with rheumatoid arthritis: A pilot study. *Alternative Therapies in Health and Medicine*, *15*(4), 24–31. Retrieved from http://europepmc.org/abstract/MED/19623830.

Calogiuri, G., Evensen, K., Weydahl, A., Andersson, K., Patil, G., Ihlebæk, C., et al. (2015). Green exercise as a workplace intervention to reduce job stress. Results from a pilot study. *Work: A Journal of Prevention, Assessment, and Rehabilitation*, *53*(1), 99–111. https://doi.org/10.3233/wor-152219.

Carlson, L. E. (2018). Uptake of mindfulness-based interventions: A phenomenon of wealthy white western women? *Clinical Psychology: Science and Practice*, *25*(3), e12258. Retrieved from https://www.mendeley.com/catalogue/uptake-mindfulnessbased-interventions-phenomenon-wealthy-white-western-women/.

Carlson, L. E., Doll, R., Stephen, J., Faris, P., Tamagawa, R., Drysdale, E., et al. (2013). Randomized controlled trial of mindfulness-based cancer recovery versus supportive expressive group therapy for distressed survivors of breast cancer. *Journal of Clinical Oncology*, *31*(25), 3119–3126. Retrieved from http://www.ksph.kz/files/journal_club/14jc/Randomized_Controlled_Trial_of_Mindfulness-Based_Cancer.pdf.

Carlson, L. E., Speca, M., Faris, P., & Patel, K. D. (2007). One year pre–post intervention follow-up of psychological, immune, endocrine and blood pressure outcomes of mindfulness-based stress reduction (MBSR) in breast and prostate cancer outpatients. *Brain, Behavior, and Immunity*, *21*(8), 1038–1049. https://doi.org/10.1016/j.bbi.2007.04.002.

Carlson, L. E., Speca, M., Patel, K. D., & Goodey, E. (2004). Mindfulness-based stress reduction in relation to quality of life, mood, symptoms of stress and levels of cortisol, dehydroepiandrosterone sulfate (DHEAS) and melatonin in breast and prostate cancer outpatients. *Psychoneuroendocrinology, 29*(4), 448–474. https://doi.org/10.1016/S0306-4530(03)00054-4.

Cartwright, T., Mason, H., Porter, A., & Pilkington, K. (2020). Yoga practice in the UK: A cross-sectional survey of motivation, health benefits and behaviours. *BMJ Open, 10,* e031848.

Chida, Y., & Steptoe, A. (2008). Positive psychological well-being and mortality: A quantitative review of prospective observational studies. *Psychosomatic Medicine, 70*(7), 741–756. Retrieved from https://journals.lww.com/psychosomaticmedicine/Fulltext/2008/09000/Positive_Psychological_Well_Being_and_Mortality__A.1.aspx.

Clow, A., & Fredhoi, C. (2006). Normalisation of salivary cortisol levels and self-report stress by a brief lunchtime visit to an art gallery by London City workers. *Journal of Holistic Healthcare, 3*(2), 29–32.

Clow, A., & Smyth, N. (2020). Salivary cortisol as a non-invasive *window on the brain.* In A. Clow & N. Smyth (Eds.), *Vol. 150. International Review of Neurobiology* (pp. 1–16). Elsevier.

Cohen, D. L., Bloedon, L. T., Rothman, R. L., Farrar, J. T., Galantino, M. L., Volger, S., et al. (2011). Iyengar yoga versus enhanced usual care on blood pressure in patients with prehypertension to stage I hypertension: A randomized controlled trial. *Evidence-Based Complementary and Alternative Medicine, 2011,* 546428.

Cuijpers, P., & Smit, F. (2002). Excess mortality in depression: A meta-analysis of community studies. *Journal of Affective Disorders, 72*(3), 227–236.

Curtis, K., Osadchuk, A., & Katz, J. (2011). An eight-week yoga intervention is associated with improvements in pain, psychological functioning and mindfulness, and changes in cortisol levels in women with fibromyalgia. *Journal of Pain Research, 4,* 189–201. https://doi.org/10.2147/JPR.S22761.

Danner, D. D., Snowdon, D. A., & Friesen, W. V. (2001). Positive emotions in early life and longevity: Findings from the nun study. *Journal of Personality and Social Psychology, 80*(5), 804–813. https://doi.org/10.1037//0022-3514.80.5.804.

Daubenmier, J., Kristeller, J., Hecht, F. M., Maninger, N., Kuwata, M., Jhaveri, K., et al. (2011). Mindfulness intervention for stress eating to reduce cortisol and abdominal fat among overweight and obese women: An exploratory randomized controlled study. *Journal of Obesity, 2011,* 651936.

Dickerson, S. S., & Kemeny, M. E. (2004). Acute stressors and cortisol responses: A theoretical integration and synthesis of laboratory research. *Psychological Bulletin, 130*(3), 355–391. https://doi.org/10.1037/0033-2909.130.3.355.

Diener, E., & Emmons, R. A. (1984). The independence of positive and negative affect. *Journal of Personality and Social Psychology, 47*(5), 1105–1117. https://doi.org/10.1037//0022-3514.47.5.1105.

Dijckmans, B., Tortosa-Martínez, J., Caus, N., González-Caballero, G., Martínez-Pelegrin, B., Manchado-Lopez, C., et al. (2017). Does the diurnal cycle of cortisol explain the relationship between physical performance and cognitive function in older adults? *European Review of Aging and Physical Activity, 14*(1), 6. https://doi.org/10.1186/s11556-017-0175-5.

Drogos, L. L., Wynne-Edwards, K., Zhou, R., Hall, S. E., Tyndall, A. V., Longman, R. S., et al. (2019). Aerobic exercise increases cortisol awakening response in older adults. *Psychoneuroendocrinology, 103,* 241–248. https://doi.org/10.1016/j.psyneuen.2019.01.012.

Evans, P., Smyth, N., Thorn, L., Hucklebridge, F., & Clow, A. (2019). Salience versus magnitude in the measurement of the cortisol awakening response. *Psychoneuroendocrinology, 103,* 249–258. https://doi.org/10.1016/j.psyneuen.2019.01.023.

Fancourt, D., Ockelford, A., & Belai, A. (2014). The psychoneuroimmunological effects of music: A systematic review and a new model. *Brain, Behavior, and Immunity*, *36*, 15–26. https://doi.org/10.1016/j.bbi.2013.10.014.

Fekedulegn, D., Innes, K., Andrew, M. E., Tinney-Zara, C., Charles, L. E., Allison, P., et al. (2018). Sleep quality and the cortisol awakening response (CAR) among law enforcement officers: The moderating role of leisure time physical activity. *Psychoneuroendocrinology*, *95*, 158–169. https://doi.org/10.1016/j.psyneuen.2018.05.034.

Field, T., Diego, M., Delgado, J., & Medina, L. (2013). Yoga and social support reduce prenatal depression, anxiety and cortisol. *Journal of Bodywork and Movement Therapies*, *17*(4), 397–403. https://doi.org/10.1016/j.jbmt.2013.03.010.

Fixsen, A., & Marie, P. Social Prescribing for stress related disorders and brain health. In: *International Review of Neurobiology,* forthcoming.

Flook, L., Goldberg, S. B., Pinger, L., Bonus, K., & Davidson, R. J. (2013). Mindfulness for teachers: A pilot study to assess effects on stress, burnout, and teaching efficacy. *Mind, Brain, and Education*, 7(3), 182–195. https://doi.org/10.1111/mbe.12026.

Foley, L. S., Prapavessis, H., Osuch, E. A., De Pace, J. A., Murphy, B. A., & Podolinsky, N. J. (2008). An examination of potential mechanisms for exercise as a treatment for depression: A pilot study. *Mental Health and Physical Activity*, *1*(2), 69–73. https://doi.org/10.1016/j.mhpa.2008.07.001.

Garde, A. H., Persson, R., Hansen, Å. M., Österberg, K., Ørbæk, P., Eek, F., et al. (2009). Effects of lifestyle factors on concentrations of salivary cortisol in healthy individuals. *Scandinavian Journal of Clinical and Laboratory Investigation*, *69*(2), 242–250. https://doi.org/10.1080/00365510802483708.

Gladwell, V. F., Brown, D. K., Wood, C., Sandercock, G. R., & Barton, J. L. (2013). The great outdoors: How a green exercise environment can benefit all. *Extreme Physiology & Medicine*, *2*(1), 3. https://doi.org/10.1186/2046-7648-2-3.

Goldberg, S. B., Manley, A. R., Smith, S. S., Greeson, J. M., Russell, E., Van Uum, S., et al. (2014). Hair cortisol as a biomarker of stress in mindfulness training for smokers. *The Journal of Alternative and Complementary Medicine*, *20*(8), 630–634. https://doi.org/10.1089/acm.2014.0080.

Gothe, N. P., Keswani, R. K., & McAuley, E. (2016). Yoga practice improves executive function by attenuating stress levels. *Biological Psychology*, *121*, 109–116. https://doi.org/10.1016/j.biopsycho.2016.10.010.

Granath, J., Ingvarsson, S., von Thiele, U., & Lundberg, U. (2006). Stress management: A randomized study of cognitive behavioural therapy and yoga. *Cognitive Behaviour Therapy*, *35*(1), 3–10. https://doi.org/10.1080/16506070500401292.

Grossman, P., Niemann, L., Schmidt, S., & Walach, H. (2004). Mindfulness-based stress reduction and health benefits: A meta-analysis. *Journal of Psychosomatic Research*, *57*(1), 35–43. https://doi.org/10.1016/S0022-3999(03)00573-7.

Hagberg, L. A., Lindahl, B., Nyberg, L., & Hellénius, M.-L. (2009). Importance of enjoyment when promoting physical exercise. *Scandinavian Journal of Medicine & Science in Sports*, *19*(5), 740–747. https://doi.org/10.1111/j.1600-0838.2008.00844.x.

Hellhammer, J., Fries, E., Schweisthal, O. W., Schlotz, W., Stone, A. A., & Hagemann, D. (2007). Several daily measurements are necessary to reliably assess the cortisol rise after awakening: State- and trait components. *Psychoneuroendocrinology*, *32*(1), 80–86. https://doi.org/10.1016/j.psyneuen.2006.10.005.

Hemingway, H., & Marmot, M. (1999). Evidence based cardiology: Psychosocial factors in the aetiology and prognosis of coronary heart disease: Systematic review of prospective cohort studies. *BMJ: British Medical Journal*, *318*(7196), 1460.

Hillier, A., Murphy, D., & Ferrara, C. (2011). A pilot study: Short-term reduction in salivary cortisol following low level physical exercise and relaxation among adolescents and young adults on the autism Spectrum. *Stress and Health*, 27(5), 395–402. https://doi.org/10.1002/smi.1391.

Ho, R. T. H., Fong, T. C. T., Chan, W. C., Kwan, J. S. K., Chiu, P. K. C., Yau, J. C. Y., et al. (2018). Psychophysiological effects of dance movement therapy and physical exercise on older adults with mild dementia: A randomized controlled trial. *The Journals of Gerontology Series B, Psychological Sciences and Social Sciences*, 1–11. https://doi.org/10.1093/geronb/gby145.

Ho, R. T. H., Fong, T. C. T., & Yip, P. S. F. (2018). Perceived stress moderates the effects of a randomized trial of dance movement therapy on diurnal cortisol slopes in breast cancer patients. *Psychoneuroendocrinology*, *87*, 119–126. https://doi.org/10.1016/j.psyneuen.2017.10.012.

Jedel, S., Hoffman, A., Merriman, P., Swanson, B., Voigt, R., Rajan, K. B., et al. (2014). A randomized controlled trial of mindfulness-based stress reduction to prevent flare-up in patients with inactive ulcerative colitis. *Digestion*, *89*(2), 142–155. https://doi.org/10.1159/000356316.

Jung, H. Y., Lee, H., & Park, J. (2015). Comparison of the effects of Korean mindfulness-based stress reduction, walking, and patient education in diabetes mellitus. *Nursing & Health Sciences*, *17*(4), 516–525. https://doi.org/10.1111/nhs.12229.

Kabat-Zinn, J. (1982). An outpatient program in behavioral medicine for chronic pain patients based on the practice of mindfulness meditation: Theoretical considerations and preliminary results. *General Hospital Psychiatry*, *4*(1), 33–47. https://doi.org/10.1016/0163-8343(82)90026-3.

Kar, S. K., Ganguly, T., Roy, S. S., & Goswami, A. (2015). Effect of Indian classical music (raga therapy) on fentanyl, vecuronium, propofol requirements and cortisol levels in cardiopulmonary bypass. *Journal of Anesthesia & Critical Care: Open Access*, *2*(2), 47.

Khoury, B., Sharma, M., Rush, S. E., & Fournier, C. (2015). Mindfulness-based stress reduction for healthy individuals: A meta-analysis. *Journal of Psychosomatic Research*, *78*(6), 519–528. https://doi.org/10.1016/j.jpsychores.2015.03.009.

Kirschbaum, C., Pirke, K.-M., & Hellhammer, D. H. (1993). The 'Trier social stress test'—A tool for investigating psychobiological stress responses in a laboratory setting. *Neuropsychobiology*, *28*(1–2), 76–81. https://doi.org/10.1159/000119004.

Klaperski, S., von Dawans, B., Heinrichs, M., & Fuchs, R. (2013). Does the level of physical exercise affect physiological and psychological responses to psychosocial stress in women? *Psychology of Sport and Exercise*, *14*(2), 266–274. https://doi.org/10.1016/j.psychsport.2012.11.003.

Klaperski, S., von Dawans, B., Heinrichs, M., & Fuchs, R. (2014). Effects of a 12-week endurance training program on the physiological response to psychosocial stress in men: A randomized controlled trial. *Journal of Behavioral Medicine*, *37*(6), 1118–1133.

Kondo, M. C., Jacoby, S. F., & South, E. C. (2018). Does spending time outdoors reduce stress? A review of real-time stress response to outdoor environments. *Health & Place*, *51*, 136–150. https://doi.org/10.1016/j.healthplace.2018.03.001.

Kubzansky, L. D., & Kawachi, I. (2000). Going to the heart of the matter: Do negative emotions cause coronary heart disease? *Journal of Psychosomatic Research*, *48*(4–5), 323–337.

Kubzansky, L. D., & Thurston, R. C. (2007). Emotional vitality and incident coronary heart disease: Benefits of healthy psychological functioning. *Archives of General Psychiatry*, *64*(12), 1393–1401. https://doi.org/10.1001/archpsyc.64.12.1393.

Kuper, H., & Marmot, M. (2003). Job strain, job demands, decision latitude, and risk of coronary heart disease within the Whitehall II study. *Journal of Epidemiology and Community Health*, *57*(2), 147–153. https://doi.org/10.1136/jech.57.2.147.

Law, R., & Clow, A. (2020). Stress, the cortisol awakening response and cognitive function. In A. Clow & N. Smyth (Eds.), *Vol. 150. International Review of Neurobiology* (pp. 187–217). Elsevier.

Law, R., Hucklebridge, F., Thorn, L., Evans, P., & Clow, A. (2013). State variation in the cortisol awakening response. *Stress*, *16*(5), 483–492. https://doi.org/10.3109/10253890.2013.817552.

Lee, J., Park, B.-J., Tsunetsugu, Y., Kagawa, T., & Miyazaki, Y. (2009). Restorative effects of viewing real forest landscapes, based on a comparison with urban landscapes. *Scandinavian Journal of Forest Research*, *24*(3), 227–234. https://doi.org/10.1080/02827580902903341.

Lewis, B. A., Williams, D. M., Frayeh, A., & Marcus, B. H. (2016). Self-efficacy versus perceived enjoyment as predictors of physical activity behaviour. *Psychology & Health*, *31*(4), 456–469. https://doi.org/10.1080/08870446.2015.1111372.

Li, A. W., & Goldsmith, C.-A. W. (2012). The effects of yoga on anxiety and stress. *Alternative Medicine Review: A Journal of Clinical Therapeutic*, *17*(1), 21–35. Retrieved from http://europepmc.org/abstract/MED/22502620.

Mackenzie, M. J., Carlson, L. E., & Speca, M. (2005). Mindfulness-based stress reduction (MBSR) in oncology. *Evidence-Based Integrative Medicine*, *2*(3), 139–145. https://doi.org/10.2165/01197065-200502030-00005.

Marmot, M. G., Rose, G., Shipley, M., & Hamilton, P. J. (1978). Employment grade and coronary heart disease in British civil servants. *Journal of Epidemiology & Community Health*, *32*(4), 244–249. https://doi.org/10.1136/jech.32.4.244.

Martin, L., Oepen, R., Bauer, K., Nottensteiner, A., Mergheim, K., Gruber, H., et al. (2018). Creative arts interventions for stress management and prevention—A systematic review. *Behavioral Sciences*, *8*, 28. https://doi.org/10.3390/bs8020028.

Michalsen, A., Grossman, P., Acil, A., Langhorst, J., Lüdtke, R., Esch, T., et al. (2005). Rapid stress reduction and anxiolysis among distressed women as a consequence of a three-month intensive yoga program. *Medical Science Monitor: International Medical Journal of Experimental and Clinical Research*, *11*(12), CR555–561. Retrieved from http://europepmc.org/abstract/MED/16319785.

Mottahedian, T. E., Sahraei, H., Movahhedi, R. S., Hajizadeh, E., & Lak, M. (2012). The effect of music on the level of cortisol, blood glucose and physiological variables in patients undergoing spinal anesthesia. *EXCLI Journal*, *11*, 556–565. Retrieved from https://www.ncbi.nlm.nih.gov/pubmed/27350774.

Noriya Uedo, M. D., Hideki Ishikawa, M. D., Kanehisa Morimoto, M. D., Ryu Ishihara, M. D., Hiroyuki Narahara, M. D., Ikuko Akedo, M. D., et al. (2004). Reduction in salivary cortisol level by music therapy during colonoscopic examination. *Hepato-Gastroenterology*, *51*, 463–465.

Nyklíček, I., Mommersteeg, P., Van Beugen, S., Ramakers, C., & Van Boxtel, G. J. (2013). Mindfulness-based stress reduction and physiological activity during acute stress: A randomized controlled trial. *Health Psychology*, *32*(10), 1110.

Ott, M. J., Norris, R. L., & Bauer-Wu, S. M. (2006). Mindfulness meditation for oncology patients: A discussion and critical review. *Integrative Cancer Therapies*, *5*(2), 98–108. https://doi.org/10.1177/1534735406288083.

Park, B.-J., Tsunetsugu, Y., Kasetani, T., Hirano, H., Kagawa, T., Sato, M., et al. (2007). Physiological effects of Shinrin-yoku (taking in the atmosphere of the Forest)—Using salivary cortisol and cerebral activity as indicators. *Journal of Physiological Anthropology*, *26*(2), 123–128. https://doi.org/10.2114/jpa2.26.123.

Park, B. J., Tsunetsugu, Y., Kasetani, T., Kagawa, T., & Miyazaki, Y. (2009). The physiological effects of Shinrin-yoku (taking in the forest atmosphere or forest bathing): Evidence from field experiments in 24 forests across Japan. *Environmental Health and Preventive Medicine*, *15*(1), 18. https://doi.org/10.1007/s12199-009-0086-9.

Pascoe, M. C., Thompson, D. R., & Ski, C. F. (2017). Yoga, mindfulness-based stress reduction and stress-related physiological measures: A meta-analysis. *Psychoneuroendocrinology*, *86*, 152–168. https://doi.org/10.1016/j.psyneuen.2017.08.008.

Paterson, D., & Murias, J. (2014). Physical functioning and mental health in older adults. In A. Clow & S. Edmunds (Eds.), *Physical Activity and Mental Health* (pp. 119–140). Champaign, IL: Human Kinetics.

Pressman, S. D., & Cohen, S. (2005). Does positive affect influence health? *Psychological Bulletin*, *131*(6), 925–971. https://doi.org/10.1037/0033-2909.131.6.925.

Puterman, E., O'Donovan, A., Adler, N. E., Tomiyama, A. J., Kemeny, M., Wolkowitz, O. M., et al. (2011). Physical activity moderates effects of stressor-induced rumination on cortisol reactivity. *Psychosomatic Medicine*, *73*(7), 604–611. https://doi.org/10.1097/PSY.0b013e318229e1e0.

Raghavendra, R. M., Vadiraja, H. S., Nagarathna, R., Nagendra, H. R., Rekha, M., Vanitha, N., et al. (2009). Effects of a yoga program on cortisol rhythm and mood states in early breast Cancer patients undergoing adjuvant radiotherapy: A randomized controlled trial. *Integrative Cancer Therapies*, *8*(1), 37–46. https://doi.org/10.1177/1534735409331456.

Rocha, K. K. F., Ribeiro, A. M., Rocha, K. C. F., Sousa, M. B. C., Albuquerque, F. S., Ribeiro, S., et al. (2012). Improvement in physiological and psychological parameters after 6 months of yoga practice. *Consciousness and Cognition*, *21*(2), 843–850. https://doi.org/10.1016/j.concog.2012.01.014.

Roe, J. J., Thompson, W. C., Aspinall, A. P., Brewer, J. M., Duff, I. E., Miller, D., et al. (2013). Green space and stress: Evidence from cortisol measures in deprived urban communities. *International Journal of Environmental Research and Public Health*, *10*, 4086–4103. https://doi.org/10.3390/ijerph10094086.

Ryan, R., Clow, A., Spathis, A., Smyth, N., Barclay, S., Fallon, M., et al. (2017). Salivary diurnal cortisol profiles in patients suffering from chronic breathlessness receiving supportive and palliative care services: A cross-sectional study. *Psychoneuroendocrinology*, *79*, 134–145. https://doi.org/10.1016/j.psyneuen.2017.01.025.

Ryff, C. D., Dienberg Love, G., Urry, H. L., Muller, D., Rosenkranz, M. A., Friedman, E. M., et al. (2006). Psychological well-being and ill-being: Do they have distinct or mirrored biological correlates? *Psychotherapy and Psychosomatics*, *75*(2), 85–95. https://doi.org/10.1159/000090892.

Schell, F. J., Allolio, B., & Schonecke, O. W. (1994). Physiological and psychological effects of hatha-yoga exercise in healthy women. *International Journal of Psychosomatics: Official Publication of the International Psychosomatics Institute*, *41*(1–4), 46–52. Retrieved from http://europepmc.org/abstract/MED/7843867.

Segerstrom, S. C., Sephton, S. E., & Westgate, P. M. (2017). Intraindividual variability in cortisol: Approaches, illustrations, and recommendations. *Psychoneuroendocrinology*, *78*, 114–124.

Seligman, M. E. P., Steen, T. A., Park, N., & Peterson, C. (2005). Positive psychology progress: Empirical validation of interventions. *American Psychologist*, *60*(5), 410.

Shi, X., Sun, X., Yao, Z., Yuan, Y., Wu, J., & Clow, A. (2018). The cortisol awakening response predicts response inhibition in the afternoon of the same day. *Psychoneuroendocrinology*, *89*, 23–29. https://doi.org/10.1016/j.psyneuen.2017.12.016.

Smith, J. E., Richardson, J., Hoffman, C., & Pilkington, K. (2005). Mindfulness-based stress reduction as supportive therapy in cancer care: Systematic review. *Journal of Advanced Nursing*, *52*(3), 315–327. https://doi.org/10.1111/j.1365-2648.2005.03592.x.

Smyth, N., Hucklebridge, F., Thorn, L., Evans, P., & Clow, A. (2013). Salivary cortisol as a biomarker in social science research. *Social and Personality Psychology Compass*, 7(9), 605–625. https://doi.org/10.1111/spc3.12057.

Smyth, N., Skender, E., David, F. J., Munoz, M. J., Fantuzzi, G., Clow, A., et al. (2019). Endurance exercise reduces cortisol in Parkinson's disease with mild cognitive impairment. *Movement Disorders*, *34*(8), 1238–1239. https://doi.org/10.1002/mds.27719.

Smyth, N., Thorn, L., Hucklebridge, F., Clow, A., & Evans, P. (2016). Assessment of the cortisol awakening response: Real-time analysis and curvilinear effects of sample timing inaccuracy. *Psychoneuroendocrinology*, *74*, 380–386. https://doi.org/10.1016/j.psyneuen.2016.09.026.

Stalder, T., Hucklebridge, F., Evans, P., & Clow, A. (2009). Use of a single case study design to examine state variation in the cortisol awakening response: Relationship with time of awakening. *Psychoneuroendocrinology*, *34*(4), 607–614. https://doi.org/10.1016/j.psyneuen.2008.10.023.

Stalder, T., Kirschbaum, C., Kudielka, B. M., Adam, E. K., Pruessner, J. C., Wüst, S., et al. (2016). Assessment of the cortisol awakening response: Expert consensus guidelines. *Psychoneuroendocrinology*, *63*, 414–432. https://doi.org/10.1016/j.psyneuen.2015.10.010.

Tortosa-Martínez, J., Clow, A., Caus-Pertegaz, N., González-Caballero, G., Abellán-Miralles, I., & Saenz, M. J. (2015). Exercise increases the dynamics of diurnal cortisol secretion and executive function in people with amnestic mild cognitive impairment. *Journal of Aging and Physical Activity*, *23*(4), 550–558. Retrieved from https://journals.humankinetics.com/view/journals/japa/23/4/article-p550.xml.

Turner, A., Smyth, N., Hall, S., Torres, S., Hussein, M., Jayasinghe, S., et al. (2020). How strong is the evidence linking stress reactivity with future health and disease outcomes?—A systematic review of prospective evidence. *Psychoneuroendocrinology*, *100*, S52–S53.

Uçar, C., Özgöçer, T., & Yildiz, S. (2018). Late-night exercise affects the autonomic nervous system activity but not the hypothalamo-pituitary-adrenal axis in the next morning. *The Journal of Sports Medicine and Physical Fitness*, *58*(1–2), 57–65. https://doi.org/10.23736/s0022-4707.16.06766-9.

Van Den Berg, A. E., & Custers, M. H. G. (2010). Gardening promotes neuroendocrine and affective restoration from stress. *Journal of Health Psychology*, *16*(1), 3–11. https://doi.org/10.1177/1359105310365577.

Vreeburg, S. A., Kruijtzer, B. P., van Pelt, J., van Dyck, R., DeRijk, R. H., Hoogendijk, W. J. G., et al. (2009). Associations between sociodemographic, sampling and health factors and various salivary cortisol indicators in a large sample without psychopathology. *Psychoneuroendocrinology*, *34*(8), 1109–1120. https://doi.org/10.1016/j.psyneuen.2009.04.024.

Vrinceanu, T., Esmail, A., Berryman, N., Predovan, D., Vu, T. T. M., Villalpando, J. M., et al. (2019). Dance your stress away: Comparing the effect of dance/movement training to aerobic exercise training on the cortisol awakening response in healthy older adults. *Stress*, *22*(6), 687–695. https://doi.org/10.1080/10253890.2019.1617690.

Ward Thompson, C., Roe, J., Aspinall, P., Mitchell, R., Clow, A., & Miller, D. (2012). More green space is linked to less stress in deprived communities: Evidence from salivary cortisol patterns. *Landscape and Urban Planning*, *105*(3), 221–229. https://doi.org/10.1016/j.landurbplan.2011.12.015.

West, J., Otte, C., Geher, K., Johnson, J., & Mohr, D. C. (2004). Effects of hatha yoga and african dance on perceived stress, affect, and salivary cortisol. *Annals of Behavioral Medicine*, *28*(2), 114–118. https://doi.org/10.1207/s15324796abm2802_6.

Wetherell, M. A., Craw, O., & Smith, M. A. (2018). Techniques for inducing stress: Problems and opportunities. In: *The Routledge International Handbook of Psychobiology*. Routledge.

Wood, C. J., Clow, A., Hucklebridge, F., Law, R., & Smyth, N. (in review). Physical fitness and prior physical activity are both associated with less cortisol secretion during psychosocial stress. *Anxiety, Stress and Coping*, *31*(2), 135–145. https://doi.org/10.1080/10615806.2017.1390083.

Wood, C., Flynn, M., Law, R., Naufahu, J., & Smyth, N. (in review). The effect of the exercise environment on the response to psychological stress: A pilot study. Anxiety, Stress and Coping.

Wood, C. J., & Smyth, N. (2019). The health impact of nature exposure and green exercise across the life course: A pilot study. *International Journal of Environmental Health Research*, 1–10. https://doi.org/10.1080/09603123.2019.1593327.

FSC
www.fsc.org
MIX
Papier | Fördert
gute Waldnutzung
FSC® C083411

Kolibri 360 GmbH
Ferdinand-Jühlke-Straße 7
99095 Erfurt, Deutschland
produktsicherheit@kolibri360.de